CW00687813

LISTEN IN

LISTEN IN

HOW RADIO CHANGED THE HOME

BEATY RUBENS

BODLEIAN
LIBRARY
PUBLISHING

For Oliver, with all my love

Published to accompany an exhibition at the Bodleian Library, Oxford

First published in 2025 by Bodleian Library Publishing
Broad Street, Oxford OX1 3BG

www.bodleianshop.co.uk

ISBN 978 1 85124 631 1

Publisher: Samuel Fanous
Managing Editor: Susie Foster
Editor: Janet Phillips
Picture Editor: Leanda Shrimpton
Cover design by Dot Little at the Bodleian Library
Designed and typeset by Lucy Morton of illuminati in 12½ on 15 Perpetua
Printed and bound in China by 1010 Printing International Ltd.
on 140 gsm Chinese Golden Sun woodfree paper

British Library Catalogue in Publishing Data
A CIP record of this publication is available from the British Library

CONTENTS

Mr Murgatroyd Yes, Mr Winterbottom, we must always study the listener – the Mr and Mrs Everymans.
Mr Winterbottom The Jones and Smiths.
Mr Murgatroyd The Robinsons and Browns.
Mr Winterbottom The Gilbert and Sullivans.
Mr Murgatroyd The Tristan and Isoldes.
Mr Winterbottom The Hengists and Horsas.
Mr Murgatroyd The Moodys and Sankeys.
Mr Winterbottom And the Darbys and Joans.
Mr Murgatroyd Cut out the Joans and let's think of the Derby. What have you backed?
Mr Winterbottom My car into a shop window. Joan a car?...[1]

The RADIO BEACON

THE ILLUSTRATED HOME MAGAZINE DEVOTED TO BROADCASTING AND THE ARTS

SPECIAL FEATURES:

FUTURE OF BROADCASTING
by Professor A. M. Low

THAT RADIO FEELING
by John Henry, Ina Janssen and John Collinson

BRITISH SCHOOL OF WATER COLOUR PAINTERS
by Isaac J. Williams

CONNOISSEURS' CORNER

QUEER CHILDREN'S PARTY
by Chrissie White

WOMEN'S CORNER

SECRET OF THE STUDIO

LIFE IN THE ANTARCTIC
by Capt. L. Greenstreet

ETC., ETC., ETC.

Vol. 1.
No. 1.

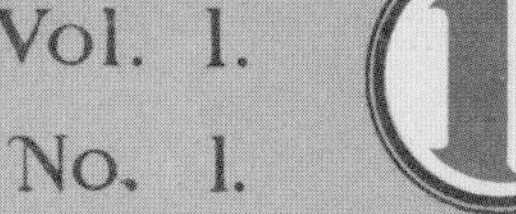

FEBRUARY,
· 1926 ·

FOREWORD

The portrait of John Reith that still hangs in the council chamber of Broadcasting House in London above a fireplace built of solid stone is forbidding. The beetle-browed first director general of the BBC glowers down on you, seemingly without pity. One of the many joys of this story of the 'wireless' by Beaty Rubens comes from remembering how misleading it is to associate Reith's name with narrow-minded authority and nothing else. Whether he always liked it or not, he was bent on a journey of adventure.

For those of us lucky enough to have had a foothold in the radio family, as Beaty has throughout her working life as an expert and imaginative producer, the whiff of romance is never far away. But it doesn't come from an imagined fantasy world. No one can read her account of the early days – as fast-moving in their way as anything we've experienced in the digital age – without reliving the thrill of a device that transformed the lives of families in the course of a generation, through a kind of magic as mysterious in its time as the very quivering of the cats' whiskers on the first crystal sets.

The Archbishop of Canterbury's wife wondered if you had to open a window to let it work properly; Ramsay MacDonald, who became prime minister two years after the BBC was founded in 1922, thought

In the early years of broadcasting, newsagents' shelves were filled with a growing range of magazines. Radio was not yet centralized, and this magazine, from 1926, was published by the *Radio Guild of the West* for Welsh and West Country listeners.

it boomed at him and wanted to be left alone – but the wireless was changing everything. By the end of the 1930s, and the coming of war, it had a place in the home as secure and proud as any Victorian piano in a middle-class drawing room – and, moreover, it was a democratizing force, bringing people together.

Anyone who has worked in radio has had this early revelation: that although every programme creates its own community – devoted, warm and often cantankerous – the medium also fosters a special intimacy. Its power is one-to-one. That voice from the loudspeaker (or, now, the mobile phone); those opening bars from an evening concert; that unforgettable burst of sports commentary; poetry and drama; the pips and the news. They are for one and for all at the same time. Beaty's story is a meticulous exploration of the early days, and the changing habits in homes across the country that came with the wireless – all the more revealing for being told from the listeners' point of view.

Reith himself hinted that the excitement of discovering broadcasting was rather like the uncovering of Tutankhamun's tomb, in the same year as the BBC was founded. Well, let's not get carried away. The advance in technology was rapid, the development of the BBC's outlook less so. Yet from the beginning – as in one of the earliest audience research projects in the working-class Barton Hill district of Bristol – everyone realized that the broadcaster, though institutionally devoted to king, church and country, was also an instrument for change. It opened everyone's ears and eyes.

There are so many gems here. The programme billing in an early Radio Times – 'If there is any News, it will be broadcast at 9.0 pm.' The tumult of the General Strike in 1926. The British Empire Exhibition a decade later, which did for radio what the Coronation would do for television in 1953. And the characters. Like Reith's might-have-been successor as director general, Hilda Matheson – brought up in a Scottish minister's manse, as he had been – who, while the first director of talks in the 1920s, was also pursuing a torrid love affair with Vita Sackville-West.

Adventure indeed. And the preciousness of this account lies in the story that stretched far from Broadcasting House, and the band of pioneers who made the wireless work, into the country beyond, in the difficult and uncertain years between the two world wars. Archives, some of them unexpected finds for Beaty, have produced a treasure trove of evidence about attitudes and a rapidly evolving way of life – from the days when primitive sets were owned by the lucky few to the time soon afterwards when every home had a radio, and mass production had made them available at a reasonable price.

You don't have to be steeped in the story of radio – although I confess I am and always will be – to enjoy the sheer verve of this revolution. As the first Christmas edition of *Radio Times* in 1923 quoted, from a contemporary song – 'So turn your earpiece to me, love / My wondrous wireless girl.' Perhaps, as Reith put it on another occasion, they were hoping to 'peer into a limitless unknown'.

Limitless, certainly. A conveyor of speech and music, thoughts and arguments and (eventually) news. As this part of the story ends, the Second World War was spawning a corps of brilliant BBC correspondents whose radio reports, often produced in the most hazardous circumstances, laid the foundations for all that we know and expect from contemporary news broadcasters. They blazed the trail on the radio – War Report was broadcast each night at 9 p.m. throughout the war, with voices from the front line. Through them, the country listened in to moments of reportage that proved indelible and are still thrilling to hear today.

This is the story of how it started. The chance discoveries, the enlightened determination, the strange gathering of figures around Reith (like him, mostly not university educated but nonetheless determined on a cultural levelling-up), that lightning change to bring into every home a loudspeaker tuned to the world beyond. Beaty Rubens has squirrelled the archives with evident excitement and the drive of a true producer. The spirit, in other words, of the wireless of long ago and the radio that we still hold dear.

James Naughtie

THE GREAT NEW WIRELESS WEEKLY

POPULAR WIRELESS

3d.

Weekly

EVERYTHING FOR THE AMATEUR

PACKED WITH PICTURES, EXPERT ADVICE, AND ALL THE LATEST WIRELESS NEWS

On Sale Every Friday

"All about Wireless"

A NEW PAPER FOR ALL

PROLOGUE

Shortly after my parents' marriage, in what I can only describe as an idiosyncratic romantic gesture, my father decided to show my mother how to build a crystal wireless receiver small enough to fit into an empty matchbox. Such sets had been a novelty in his boyhood, but time had passed and he found that his hands were no longer as deft as they had once been, so it was my mother who inserted the simple components and connected them up. As the wires sparked into life and the crackling static cleared, each put an ear to a single cup of the headphones and together they heard the remote but distinct sound of human voices.[1]

I grew up believing that everyone's father tinkered with wireless sets as mine sometimes did. Long after he died, I found a faded foolscap envelope containing a neat sheaf of receipts made out in his name from a large radio and electrical company in Liverpool where he grew up. In his teens, during those exhilarating days of the 1920s' wireless craze, he had clearly been saving his pocket money for the weekly instalments to buy the individual components and build his own first valve set. My mother, who was rather younger, was born within the BBC's first year, so she was a true child of the radio age – we might now say a 'wireless native'. In the same way as a later

Popular Wireless was launched in mid-1922, a few months before the opening of the British Broadcasting Company (as it was then called).

generation would take television for granted, and a later one still the internet, she grew up treating wireless like an item of furniture that had always been in the room, an outside source of speech, music and entertainment which could be accessed simply by pressing a switch, turning a dial. She and her sister would listen to *The Children's Hour*, and I know that her parents, my grandparents, would work their way through the *Radio Times* to mark in advance the programmes they wanted to hear, because I saw them do so in my childhood in the 1960s. I even remember the cloth slipcover on which my grandmother had embroidered a dainty image of a cottage at the end of a path in an English country garden bright with roses, lavender and hollyhocks.

I have been a radio producer for over thirty-five years and I like to think that I was conceived with a special passion for radio, but I actually believe most people have similar connections stored away in the memory banks of their own families. I am also of the view that exploring the era when wireless first arrived in British homes and examining its impact are about far more than nostalgia.

Like many people, I found time during the Covid lockdowns to do some research into my family history. Concurrently, I was reviewing for professional reasons what happened to be a similar chronological span. As 2022 approached, there were many discussions among my colleagues about how we, in the radio division, should mark the centenary of the BBC. Most were about broadcasting, broadcasters, broadcasts back in that era, sometimes called the 'golden age' of radio, between 1922 and 1939. Radio producers constantly give thought to how their programmes will be received or consumed, so it seemed odd that few of these discussions mentioned listening or the listeners. The statistics alone were astounding: in 1922 a little under 150,000 people had the wherewithal to listen to wireless; by 1939 the number had risen to almost 34 million.[2] In under two decades, wireless had brought an incomparably wider world into homes. How did people feel, though, when they went out and bought their first set or acquired the components to build one? What had it been like for individuals and families to integrate this newcomer, to gather and listen by their own firesides in those pre-central-heating, often pre-electricity, days?

Had the changes been entirely positive or did they miss their former domestic lives? What were their thoughts and reactions when they started to listen to the content of the first ever broadcasts – the music, talks and news bulletins – and to understand that countless other strangers across the country were sharing the same experience at the same time? Had the arrival of broadcasting in fact been as dramatic as I was assuming? Did it revolutionize their lives or might it just have reinforced the old ways?

✧

The literature on early radio and the BBC is immense, but writing from the listener perspective is rare and it's not hard to see why. Grassroots contemporary testimony is scant and scattered, as are slightly later first-person reminiscences. Radio listening is by nature so fugitive and began to be taken for granted so quickly. There are also problems in generalizing about any kind of experience within a country made up of four nations, each with its own distinct culture and heritage, which made up a combined population of over 47 million by the outbreak of the Second World War. On the other hand, as any radio producer will tell you, authentic individual experiences, told verbatim and in real time, can articulate the experiences of many.

I started by gathering such accounts as I could find, both in contemporary books about broadcasting and in later scholarly works, which had either been brilliantly researched or written close enough to the times to include some first-hand testimony. I was amused (and sometimes horrified) as I sorted through the copious cartoons which had been published in magazines and newspapers, beginning to understand that their number alone was evidence of how broad and deep the cultural impact of wireless had been, but also learning to parse with caution complex stories told in apparently simple drawings. I looked at the extraordinary resource represented by the *Radio Times*, every edition available online,[3] though I knew that I needed to tread carefully with a magazine which was, after all, billed as 'the official organ of the BBC'. I handled with particular wariness the weekly page titled 'What the Other Listener Thinks', because I knew, of course,

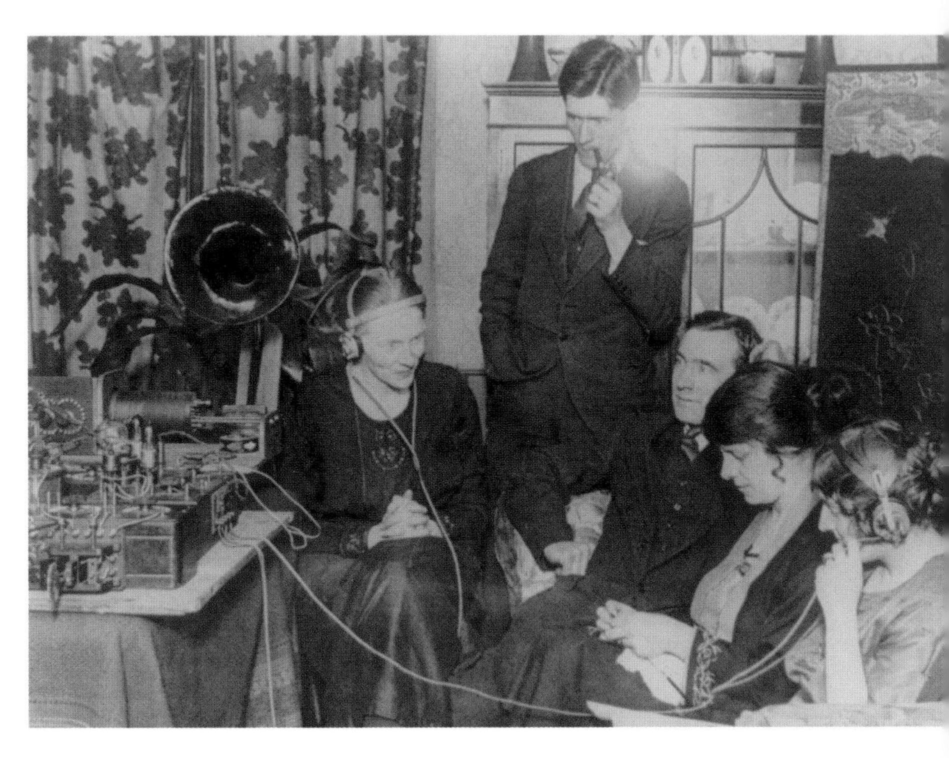

Originally published in an American newspaper in November 1923, this photograph shows an English family sharing the novel experience of listening in their own home to the results of a general election.

that somebody back then had filtered the immense postbag, and that writers and senders of letters are, by definition, self-selecting. I trawled some of the numerous other magazines and periodicals – the specialist ones intended for wireless enthusiasts but also those which simply featured weekly columns for the growing numbers of curious members of the general public.

I found, to my great delight, that the extraordinary John Johnson Collection of Printed Ephemera in the Bodleian Library in Oxford included a dozen boxes related to wireless: early advertising brochures which showed me how the manufacturers wanted radio to be perceived; BBC pamphlets which actually instructed early listeners how to listen; cigarette cards, games and puzzles; radio licences and certificates to prove membership of the Radio Circle, which led me, in turn, to look elsewhere for information about its direct commercial rival, the League of Ovaltineys. The Marconi Archives – its documents held in the Bodleian's Weston Library, its objects in the History of Science Museum in Oxford – contained invaluable material and even provided me with an opportunity to handle a swordstick. I turned to Mass Observation diaries for later chapters. I pounced upon a number of casual mentions of wireless in the literature of the times: Virginia Woolf, George Orwell and E.M. Delafield – her journalism and semi-fictional diaries effervescent as brut champagne. Occasionally, I allowed myself to include later writing when I felt it contained authentic traces of the period – a poem by Seamus Heaney, a play by Brian Friel, a novel by Penelope Fitzgerald, a compendious family portrait by Maria Stepanova.

In spite of this haul, I still felt rather distant from those early listeners themselves. Then, just as I was beginning to flag, I made a discovery late one afternoon which excited me so much that I had to leave the library and go downstairs to the cafe for a cup of tea and a slice of carrot cake. It enlarged my evidence base and my understanding beyond measure. For a (significantly) long time in its early days, the BBC did not undertake audience research. When the first head of a new Listener Research unit, Robert Silvey, was eventually appointed in 1936, one of his early decisions was to commission two remarkable women to conduct a survey titled *Broadcasting in Everyday Life: A Survey of the Social Effects of the Coming of Broadcasting*.[4] They were Hilda Jennings and Winifred Gill, and the focus of their project was a working-class district of Bristol called Barton Hill where they already had connections. Unfortunately, the pamphlet they produced didn't get much press coverage at the time because the publication date happened to be in the week when Britain entered the war. It was far from unknown to historians, though, and I knew about it myself and had seen a small number of quotations cited in other works. I was also aware that rather few copies had survived, so I ordered one up to take a look. Because I had so far done all my research in the Special Collections reading room in the Weston Library and understood how it worked – the friendly silence, the lovely helpful librarians, the free provision of freshly sharpened pencils – I

BROADCASTING IN EVERYDAY LIFE

A SURVEY OF THE SOCIAL EFFECTS OF THE COMING OF BROADCASTING

Conducted on behalf of the BBC by Hilda Jennings and Winifred Gill of the Bristol University Settlement

Price One Shilling

BBC

Winifred Gill seems to have used her spiral-bound notebook in the field and then made a fair copy of excerpts in an exercise book. The pamphlet published in 1939 featured some but by no means all of her interviews.

THE SPIRAL

REPORTER & STENOGRAPHERS NOTE BOOK

MANUFACTURED UNDER PATENT 219547

300

61

Dr. de Lacy O'Leary.

services. No effect noticed.

R.C. services are nonconformist.

Mr. ? Morse —

Many wife I know. She had a real grievance against him, he did not give her a fair share of his wages & keep rare on. But if ever she started to talk about it, he put the wireless on so loud it drowned her. And when he went out he'd say — "Now I've tuned in to a foreign station & I shall expect to find it there when I come in — so don't you touch it". So she couldn't have it on, neither while he's away. But one week-end he went away to a conference, & as soon as he was gone, she said to her boy & girl "Now we'll have the wireless", but she found he had disconnected it.

She's left him now.

requested the copy held there rather than the one in the general library stacks. It was part of Winifred Gill's archive, which had been left to the Bodleian after her death. During her full and varied life she had been involved with the Bloomsbury Group, and with early social work, and these are what much of the archive covered, but there was just one dove-grey box file, catalogued 'Radio Audience Research'. When I opened it, I found the pillar-box red pamphlet ('One Shilling'), printed in handsome, official BBC font. Beneath this, though, among a few other papers, I discovered two books – a red one like a school exercise book and a green ring-bound reporter's notebook – in which she had made extensive notes, in her elegant handwriting, of the conversations she had conducted with the women and men of Barton Hill. I cross-checked the manuscript notes with the pamphlet and began to identify the material which had been used for publication. Much had not, sometimes because of a shortage of space, sometimes because it didn't altogether meet the brief of the project, or, in just a few cases, because it revealed too disturbing a picture of domestic life. Again, any radio producer will tell you that a good edit is essential, but that character and colour often emerge unexpectedly in a slightly longer or more detailed extract.

Barton Hill is not a large area and theirs hadn't been an extensive research project in the first place, but now I had far more original testimony: Mr James the irrepressible Welsh grocer, Mrs Britton the chatty newsagent, politically astute Alderman Hennessy, proud autodidact Mrs Morse, modest, thoughtful Miss Vile (my personal favourite), Eileen and George, Joyce and Les, Eunice and Ken – people with real names and identities. Not only could I read their views on wireless; I now had a clearer idea about related questions which seemed important to them and to Gill and Jennings at the time. What to do with a now redundant piano in the front parlour? Why listening to boxing matches relayed from America in the middle of the night held such appeal for entire families? How to interpret the significance of an errant husband choosing to abandon keeping pigeons in favour of breeding canaries? The research referred to the arrival of wireless in this area over the previous ten years, so these voices mostly spoke of listening experiences beginning in around 1928, but it was clear that a few went back further. The contents of the notebooks in turn enabled me to extrapolate and think more fully about other areas of the country, different classes and types of people, and the impact of the arrival of wireless on them all.

In telling this story, I follow a chronological path, beginning each chapter with a major public event. In each chapter, though, I explore a theme, and for this I often spread my gaze across the entire period: how listening to wireless encouraged a sense of shared nationhood in spite of the thorny British issues of class division and regional difference; what listening to a medium originally owned by men meant to women, to children and to family dynamics in the home; how the BBC was surprised to discover that the public at large did not universally share its passion for 'uplift', and almost lost the plot, but found it again, just in time, through laughter; the particularly contemporary questions of accuracy, reliability and trust in news broadcasts. I start with what is perhaps rather a surprising degree of Victorian 'pre-history', because the more I learnt about early radio listening, the more I understood the importance of that background. Throughout, although this book is not in any way a history of the early BBC (or of its later competitors), I sketch some key developments in the first seventeen years of wireless. What I have tried to do, above all, is to trace the growing recognition that radio was here to stay, that it represented a cultural form in its own right, and that it was changing every home across the country.

As I wrote, I frequently revisited the history of my own family, asking myself questions about their radio listening and the impact it must have had on their lives. I thought about my eccentric circle of unmarried great-great aunts, who had grown up in the East End of London and never lost their cockney accents, although their great-niece, my mother, spoke pretty much like a BBC announcer. I recalled affectionate reminiscences of the toy shop and dolls' hospital they had later run in Acton High Street, then on the western edge of London, and of their throwing an impromptu party in the room behind the shop when my grandfather came to visit before setting off in uniform for France. I reckoned this meant they had owned a gramophone player by 1915, which led me in turn to consider when they had acquired a wireless and where they had placed it in their crowded home. There was also the question of which of them it had been – Sarah, Fanny, Rose or Minnie – or perhaps one of the two unmarried great-great-uncles – Ben or Percy – who had acquired the knack of twiddling the cats' whiskers of a crystal set or assumed weekly responsibility for taking the cumbersome batteries of a valve set to be recharged. Three of them had had disabilities, and I thought of how the arrival of radio must have enhanced their lives in particular. Since the shop was also a newsagent, I wondered how they had felt about the growing influence of wireless news. I imagined which of my aunts and uncles had attended 'wireless parties' when they were a thing, or danced to wireless-themed songs. In a shallower moment (while getting my hair cut) I wondered if any of the aunts had been tempted to experiment with a wireless-inspired hairstyle. I contemplated whether my upwardly mobile London grandfather, Henry, had been an early wireless adopter soon after he and my grandmother, Lily, married in 1920: which model he might have chosen, how radio-listening had helped shape and endorse his hard-earned sense of success, with what intense anxiety his family, the owners of a small industrial workshop, must have listened to the *News* during the General Strike of 1926.

I tried to guess what Nellie, my grandmother on the other side, in her (sadly) downwardly mobile Liverpool setting, might have thought if she heard early broadcast talks on topics such as 'Women in Public Life, 'Women in Other Lands', and – the mother of two sons and a daughter – 'Openings for Girls with a University Education'.[5] I contemplated whether she had been making a quiet statement by retaining her distinctive Sunderland accent when, eventually, the family moved south, and first her daughter and later her son (my father) married into the same, more prosperous and successful, family.

Since my family is Jewish, I tried to imagine how it felt to acknowledge the close association of the early BBC with the Church, which led me to think again about diversity and assimilation and the role that the media can play in these issues. When I leafed through cartoons and read the Mass Observation reports of the Coronation of George VI in May 1937, I wondered if any of them had actually discussed whether it was respectful to be eating while listening to the broadcast of the more sacred moments of the ceremony, or necessary to stand up when the national anthem was played on the wireless.

I particularly enjoyed recorded snatches of the comedians, whose impact was so great in the years leading up to the Second World War, because I knew that my parents and their friends must surely have listened to them and shared their catchphrases: 'I don't mind if I do...', 'Can I do you now, sir?', 'Light the blue touchpaper and retire immediately!', and even 'What did Horace say, Winnie?' (though, knowing my family's instinctive loyalty to the BBC, perhaps not). When I looked over a letter to my father from an equally enthusiastic hobbyist friend who mentioned how remarkably clear reception had become, even in the remote location from which he was writing on the eve of war, I thought of the massive technical advances which had been made during these first seventeen years.

I mention such impressionistic personal musings because, just as they have sometimes helped me to reanimate the historical material on which this book is based, so I hope they will enable readers to do the same for themselves. Some of the questions which were asked at the time about this great cultural gift we are asking afresh a hundred years later. What was at first called 'wireless', and then became known as 'radio' (I use the terms interchangeably), once again, in the twenty-first century, has a new name: 'audio'. 'Downloads', 'listen again', 'catch-up' and, above all, the sensational new world of podcasting, simultaneously share many characteristics with and are vastly different from what we now call 'linear' or traditional radio. In other words, this is a book about more than a thrilling era in the not-so-distant past. As if striking the matches emptied out of that box to make a miniature wireless set all those decades ago, I hope it might kindle some debate about how we listen (or watch or consume) today – together, alone, at home, on the move, to the same or vastly different output – and why it matters.

The Radio Times, December 21, 1923

THE CHRISTMAS NUMBER

RADIO TIMES

"JUST A SONG AT TWILIGHT"

6D.

Vol. 1.—No. 13.

ONE

WHAT IS HOME WITHOUT A RADIO?

> [I]t is confidently stated that a single wire would carry the same sound over the whole United Kingdom, if not beyond the seas ... the Pleasure Telephone opens out a vista of infinite charm which few prophets of today have dreamed of, and who dare to say that in twenty years the electric miracle will not bring all the corners of the earth to our own fireside?
>
> Arthur Mee in the *Strand Magazine*, 1898[1]

In 1935 *The Times* newspaper published a special supplement on the theme of 'Home'.[2] One of the writers commissioned was E.M. Delafield, whose wry semi-autobiographical *Diary of a Provincial Lady* had recently become a bestseller and has never been out of print since then. Contrasting the Victorian domestic scene of her own childhood with contemporary home life, Delafield commented, in passing, that wax flowers, beadwork and sentimental songs had given way to painted cocktail sets, legless tables and American jazz, but she was unequivocal about the most significant change: 'If I were asked to name a symbol of modern home life', she wrote, 'I should choose the wireless.' In a few strokes, she painted a memorable picture of how the new technology was now transforming the lives of old and young, rich and poor, educated and less educated, rural, urban and suburban people:

The iconic first ever Christmas *Radio Times* and first colour cover declare the promise of radio to enhance the home and family life within it.

> All over the country, in every house, sooner or later the well-known discussions arise. Jazz or no jazz. Symphony concert or variety. Henry Hall as a background to conversation: conversation as an interruption to Henry Hall. Absolute silence for the News, or the News ignored until next morning's paper.

The impact of the arrival of wireless is sometimes compared with that of the printing press or of the steam engine, but an understanding of massive public change can all too easily overshadow or sideline personal dimensions – the shifts in household habits, the awakenings of new tastes, the alterations and adaptations of attitudes by individual people, each in their own home. Like early broadcast radio itself, very little remains of these earliest reactions, which is why every first-person testimony is so valuable. The naivety of Edith Davidson, daughter and wife of an Archbishop of Canterbury, inquiring in 1923 whether it was necessary to leave a window open when listening to the new device,[3] set against the tech-savvy Mr James, a Bristol grocer, explaining in 1938 how wireless listening has come to punctuate his day: 'I wouldn't miss the News. I've even neglected the bacon machine for the News.'[4] Labour leader Ramsay MacDonald's initially Garbo-like stance: 'so far as I am personally concerned, I am absolutely indifferent to all this booming, whether it is done by hostile or friendly people. I prefer to be left alone',[5] in contrast to precise Miss Vile, in Bristol again, explaining what wireless has come to mean to her: 'I only listen to what I want. I only listen in leisure. I wouldn't be without it… One forgets a lot, but it's not all lost. It may not be substantial knowledge, but it beautifies your life.'[6] It beautifies your life: truly an accolade which any radio producer would be delighted to hear, then or now.

In this first era, between 1922 and 1939, everything that radio listeners now take for granted was novel: a mysterious, sometimes alarming form of new technology and an unfamiliar gadget in their homes; voices and music arriving though the 'ether'; companionship; a sense of a unifying – or dividing – experience; the *News*, the 'pips', the bongs of Big Ben, the piteous SOS appeals; the potential to eavesdrop on a wider world and understand it a little better.

Queen Victoria's funeral procession enters Paddington Station on 2 February 1901, observed by crowds. Wireless broadcasting was to transform the national experience of royal deathbeds and funerals by the time her grandson, George V, died in 1936.

It's a story which begins in the nineteenth century, when most early listeners had themselves been born. The first few generations of 'wireless native' children were in time to play an important role, but this chapter reaches some way back, to the technologies which had seemed revolutionary to their parents and grandparents. Even though wireless was to transform – some might now use the term 'disrupt' – Victorian culture and supersede Victorian technologies in such deep and far-reaching ways, it was in nineteenth-century soil that the newcomer took root.

When Princess Alexandrina Victoria was born at Kensington Palace in London on 24 May 1819 it took over fifteen hours for the happy news to be carried to Bristol, and fifty hours to reach Edinburgh, even though the mail coaches changed horses every 10 miles and the messengers didn't stop for meals. By the time she died on 22 January 1901 – now Queen Victoria of Great Britain and Ireland, Empress of India – only seconds were needed for the announcement to be transmitted across the country, the British Empire and the entire world.[7] The queen had died at her rural retreat on the Isle of Wight, and the mournful journey of her coffin by royal yacht, gun carriage and train to Windsor, where the funeral would be held, was a slow one in an increasingly fast-paced world. The development of the railway, the telegraph and the telephone had pulsed through the eighty-one years of Victoria's lifespan like an electrical charge through copper wire. Then, just five years before her death, in 1896, a dynamic young Irish-Italian electrical engineer called Guglielmo Marconi had arrived in Britain and promptly patented a radically new communication system. As its name makes clear, this was wire*less*, and even Marconi himself did not comprehend its potential breadth and reach. In particular, he failed to foresee that the key function of wireless was to be the then unknown medium of broadcasting – making possible, for the first time ever, the transmission of information and entertainment directly into homes.

Of course, no two households are or were the same, but, just as a starting point, this cosy cameo of a Victorian family of modest means provides a glimpse into an evening at home in the pre-wireless world:

> Mrs Meyrick's house was not noisy: the front parlour looked on the river, and the back on gardens, so that though she was reading aloud to her daughters, the window could be left open to freshen the air of the small double room where a lamp and two candles were burning. The candles were on a table apart for Kate, who was drawing illustrations for a publisher; the lamp was not only for the reader but for Amy and Mab, who were embroidering satin cushions.[8]

This is from George Eliot's novel *Daniel Deronda*, which was published in 1876 – the year the telephone was patented. It takes a leap of the imagination to picture an utterly different scene in which Kate, Amy and Mab are now rolling back the threadbare Persian carpet and dancing – or perhaps watching their children dance – to music being played on the wireless set on the mantelpiece in the parlour, but this was to become entirely possible within their (fictional) lifetimes.

Among the Meyrick family's few possessions is a piano, and the lyrics and culture of the most popular form of Victorian home entertainment – parlour music – provide a surprisingly helpful bridge between the pre-wireless and wireless worlds. It's also important because the early promoters of broadcasting would cleverly latch on to this, recognizing that they could encourage take-up and allay mistrust and fears by summoning up the familiar shared vocabulary of hearth and home. The most famous and popular parlour song of all was 'Home Sweet Home', which had first been performed in 1823 and never lost currency:

> Mid pleasures and palaces though we may roam
> Be it ever so humble, there's no place like home...

Such songs often painted a reassuringly timeless picture of the Victorian home. Well-loved household objects such as an old oak rocking chair, a spinning wheel or a faithful grandfather clock were fondly recalled by the light of the flickering hearth, as if to offset the radical speed of social and physical upheaval. In fact, the late Victorian parlour was itself gradually becoming host to new gadgets.

Almost half a century before radio, one particular item began to feature on many household wish lists. In 1877, aged just 30, the prolific American inventor and businessman Thomas Edison had patented the phonograph, a device which incised sound vibration waveforms onto rotating wax-coated cardboard cylinders. In their neat wooden boxes, phonographs (and soon gramophone players) could be surprisingly small devices, though it was the emblematic flaring horn, initially crafted from papier mâché but soon made in engraved and painted copper or elegant panels of jointed wood, which proudly announced their presence to visitors.

The phonograph did not just bring outside voices into the home; it was also to nudge forward another, less tangible, cultural shift. As

This handsome (and expensive) oak horn loudspeaker was manufactured by the Amplion Company and introduced in around 1924. Advertisements promised 'extreme clarity' and a 'wonderfully natural tone to the enjoyment of all listeners-in'.

mid-Victorian home entertainment had increased in popularity, the debut performances of so-called parlour songs would often actually take place on the concert platform. A famous (and handsomely paid) singer could vastly increase sales of sheet music for domestic use. Now, as Edison's invention improved in sound quality and became more widely affordable, an international cast of top performers began to shift their celebrity endorsement from live shows to recordings. Before long, Enrico Caruso, Adelina Patti and Feodor Shaliapin had all been won round to the phonograph. It is the Australian lyric soprano Nellie Melba, though, who is a key player in this story. After extensive wooing, she eventually agreed to an experimental studio session, and – with an eye to the hugely lucrative popular market – her first releases included parlour songs, familiar art songs and easier opera arias. Now her massive fan base could sit in their own homes and listen to her voice again and again. With this following, it's no surprise that Nellie Melba was soon to play a key role in the launch of radio. When that time came, her first song was, of course, 'Home Sweet Home'.[9]

Before the arrival of the phonograph, when a family or group of friends had gathered around the piano for an evening of amateur music-making, there might possibly have been one among their number who had heard the songs, which they were themselves now singing, being performed in public by a professional. Mostly, though, they would be creating their own interpretation from the sheet music. Recording technology changed this forever. The formidable Russian writer Maria Stepanova vividly describes this change in her account of her own family's long history, *In Memory of Memory*: 'Caruso and Shaliapin now strode into the family parlour. In the new century, people didn't sing, they sang along and they knew the music not from the page, but from the voice, the raw and irresistible original.' Stepanova concludes with a reference to Nipper, who served as a model for the 1898 painting by Francis Barraud which was to become one of the most iconic trademarks of all time: a small white terrier-mix with soft brown ears, who stands alert, his head cocked, listening attentively to the familiar sound emanating from an outsize gramophone horn: 'Music, like much else, became an instance of deferring to authority, *his master's voice*...'[10] The early BBC was soon to lay down strict rules about how announcers spoke on air and about what they said (and did not say). Through gramophone records, listeners were becoming accustomed to this idea of an authoritative – one might almost say authorized – version, 'his master's voice', which might sometimes prove reassuring and sometimes problematic.

⁂

Many households soon boasted a phonograph but each played music according to their own taste and personal schedule: one might opt for

Nellie Melba singing an aria from Verdi's *La Traviata*; another might prefer music-hall star Marie Lloyd's rendition of the 'Piccadilly Trot': this was not a shared experience. Nothing fully prepared the public for the wireless craze of the 1920s, but *the* key innovation of the Victorian communications revolution – the telegraphy network and its offspring, the telephone – had already begun unifying the country into a powerful if invisible network.

Before there was radio (initially called 'wireless telephony') there was wireless telegraphy, and before there was wireless telegraphy there was wired or electric telegraphy. After Samuel Morse's 1837 invention, news could, for the first time, travel faster than a messenger on horseback, by means of electrical pulses through wires. Simultaneously challenging horsepower, the first modern railroad had opened in 1830, carrying people and freight, initially between Liverpool and Manchester, and, before long, across the entire country. As the sight of wires connecting the transmitting and receiving instruments of the telegraph, often strung for convenience along railway lines, became familiar, so the word 'wire' entered into common usage for the message itself (and has only recently become obsolete). A celebratory poem written in tribute to Morse in 1872 gives this process a mythological grandeur:

> One morning he made him a slender wire,
> As an artist's vision took life and form,
> While he drew from heaven the strange, fierce fire
> That reddens the edge of the midnight storm;
> And he carried it over the Mountain's crest,
> And he dropped it into the Ocean's breast;
> And Science proclaimed from shore to shore,
> That Time and Space ruled man no more.[11]

It may not be the finest poetry, but the final line, about the incidental changes to people's perception of space and time, highlights how everyday life was changing. Once telegraphy and the railways began to gather momentum, it quickly became clear that something had to be done to unify local time into a single, nationwide clock. By 1847 the Railway Clearing House had recommended that every railway company in Britain should use London time in all timetables and at all stations. When listeners later started setting their watches to the chimes of Big Ben or the sound of the 'pips' on the wireless, they were replacing the now familiar but external concept of 'railway time' with a broadcast timekeeper which actually came into each home across the country.

Since so many people listen to radio output on mobile phones today, it should perhaps not come as a surprise that telephone networks were the means of distribution for a failed but visionary early experiment in what was to become broadcasting. As use of telegraphy escalated, experiments carried out to expand its bandwidth led to a new discovery – the telephone. By 1878 Alexander Graham Bell had demonstrated the gadget to Queen Victoria herself. It was an overnight sensation. The telephone not only extended the marvellous potential of the electric telegraph but, as an advertising flyer at the time enticingly declared, it wiped out physical distance at a far more individual and social level than the telegram: 'Persons using it can converse miles apart, in precisely the same manner as though they were in the same room.'[12] Less than a decade after its invention, a wildly ambitious commercial exploitation of the technology was in development. The Electrophone is almost entirely forgotten but it deserves more than a footnote in the story, not least as a precursor of a growing twenty-first-century trend for what we now in fact call 'narrowcasting'.

Visitors to the International Electricity Exposition in Paris back in 1881 had been impressed by the inventor Clément Ader's demonstration of what he called the Théâtrophone. Ader had set up telephone microphones across the front of the stage at the Paris Opéra, forty to the left and forty to the right, to create what would later be called a stereo image. He then ran wires from the theatre, through the Paris sewers, to the exhibition space 2 kilometres away. Here, twenty receiving units were made available to excited visitors, who could listen on headphones. By the new decade Ader had found venture capitalists willing to back his media start-up.

The wealthy could subscribe to the new technology by installing a second telephone line at home and paying an annual fee. All they then had to do was contact the central switchboard as if they were making a private call, but the operator would instead connect them up to a choice of theatres or concert halls. Even one telephone line was a luxury at the time, never mind two, so this was very much a novelty for the rich, though anyone could listen to a concert or play if they had a half-franc coin to spare and could find one of the public slot machines which were soon popping up in cafés, brasseries and department stores.

Keen to encourage take-up, advertisers quickly saw the double benefit of targeting women, both those who were venturing out in public and traditionally inclined women who were still opting to remain largely at home. Perhaps the most iconic advertising image of all was created by the painter and lithographer Jules Chéret, whose familiar belle époque images have given him the name 'father of the modern poster'. Below lipstick-red letters advertising Théâtrophone a young woman faces the onlooker in three-quarters profile. Her canary-yellow dress is cinched in at her wasp-like waist and hangs in sharp folds across her bottom and up towards her breasts. Her slender arms extend diagonally upwards from her elbows to her dainty little hands, which are clasping against her ears a set of headphones connected by two bright-red flexible cords to a neat box on a shelf. Her smiling lips are bright red too, but it is the look in her eyes that is most striking, both saucily come-hither and absorbed by some distant elsewhere all her own. This idea that audio could transport listeners to a new dimension – as if by polishing a magic lantern or stepping onto a magic carpet – was to be promoted as a key attraction of radio.

By the time the sobriquet *Gai Paris* was reaching its turn-of-the-century zenith, a version of the new service had arrived in Britain. Even after a rebrand, advertising copywriters once again pitched what they called the 'Electrophone' squarely at women. An 1896 flyer

Listening to live music and voices transmitted via telephone wires to headphones became a glamorous novelty with the arrival in France of the Théâtrophone in the 1890s.

THÉÂTROPHONE
JChéret

July 1896

THE ELECTROPHONE.

The Human Voice and Music conveyed along Wires direct from the Performance to the Ear of the Listener, with all the most delicate graduations of tone and emphasis.

"I COULD NEVER HAVE BELIEVED IT POSSIBLE!"

For particulars apply to

promised 'The Human Voice and Music conveyed along Wires direct from the Performance to the Ear of the Listener, with all the most delicate graduations of tone and emphasis.'[13] The female figure is more demure in pose than her French counterpart, but she is dressed in the latest fashion – massive puffed sleeves and a deep ruffled collar and pleated bodice exaggerating the painfully tiny circumference of her waist. One hand holds a decorative ostrich feather fan, the other the rod of a lorgnette-style set of headphones. (This is an early appearance of one of today's 'must-have' accessories – a set of headphones.) Her eyes have a far-away stare, calling on the viewers to judge what she might be listening to from among the (reassuringly improving) halo of options which radiate in an elegant font around her head: 'Pulpit', 'Politics', 'Drama', 'General News', 'Opera'.

Canvassers for the company would go from door to door in wealthy areas, delivering flyers like this and urging prosperous householders to subscribe to the service. The directors of the company enthusiastically described the endless and varied amusement that the Electrophone would bring to the family circle: 'On wet evenings or when friends are gathered together the Electrophone furnishes a never-failing source of pleasure and recreation.' There was also talk of an almost-unimaginably exciting future when live trials, scientific lectures and even debates in the Houses of Parliament might be on offer to listeners. Before long, Queen Victoria became a fan, listening from Windsor Castle to a special performance by the actor Beerbohm Tree in celebration of her Diamond Jubilee. But if George Eliot had lived longer and written a sequel to *Daniel Deronda* in which Kate, Amy and Mab appeared as grown women, they would not have been subscribers. They might well have become adept at dropping a soft steel stylus onto a gramophone record or speaking into the mouthpiece of an elegant candlestick-style telephone, but the Electrophone was the equivalent of the latest high-tech gadgets which appear in spending supplements of top-end newspapers today. Even so, press coverage was

Advertising of the Electrophone in Britain in 1896 was rather demure. The lorgnette-style headphones held by this woman suggest a domestic delicacy quite different from the racy, girl-about-town body language of her French counterpart.

widespread and excitable, beginning to frame many questions which would soon be applied to wireless. What would it provide? Who would it be for? Might its function be (say) to inform, educate and entertain? Where and how might people listen? Then there were the ever-important issues of interior design. Would it become necessary to shift round furniture in the parlour? Was it possible – surely not? – that the newcomer might one day usurp a flickering fire as the focal point of any home?

One of the journalists who became entranced by the potential of the Electrophone was Arthur Mee. He was later to make a name for himself with his hugely popular *Children's Encyclopædia*, but in 1898 he wrote a long article in *The Strand Magazine* about what he called the 'Pleasure Telephone',[14] and the service he understood was already being provided by the 'Telefon Hirmondo' company to thousands of subscribers across Hungary.

With a breathless enthusiasm which disingenuously blends fact and futurology, Mee describes how each subscriber could opt into the daily output: news and a newspaper roundup each morning; information from the stock and corn exchange; theatrical, art and science notices; ecclesiastical intelligence; foreign, provincial and sporting information; society and political updates – all 'brought to one's own fireside, without the trouble of running into the streets for the papers'. It's not at all clear how much of this ever really materialized, but there were other writers who believed that the new medium might make a serious contribution to radical political change.

The source behind Mee's excitement was an influential American novel which had been published a decade earlier. The premiss of Edward Bellamy's utopian *Looking Backward 2000–1887* involves a Bostonian who wakens from a sort of fairy-tale sleep in the year 2000 to find that industry has been nationalized, wealth has been fairly distributed and class divisions have been eradicated, along with war, poverty, prostitution, crime, money and taxes. For Bellamy, the proposed listening service would provide an egalitarian lifestyle for all. Albeit in a considerably more muted form, much (though not all) of his fantasy description would soon apply to the reality of wireless:

fitted in our houses just as gas or electricity is now. It will be so cheap that not to have it would be absurd, and it will be so entertaining and useful that it will make life happier all round and bring the pleasures of society to the doors of the artisan's cottage. That indeed will be the unique feature of the Pleasure Telephone. It will make millions merry who have never been merry before, and will democratize, if we may so write, many of the social luxuries of the rich. Those who object to the environment of the stage will be able to enjoy the theatre at home, and the fashionable concert will be looked forward to as eagerly by the poor as by their wealthy neighbours. The humblest cottage will be in immediate contact with the city, and the 'private wire' will make all classes kin.[15]

It's frustrating – and, of course, telling – that not a single British subscriber recorded their actual experience or impressions of listening to the Electrophone. One major French cultural figure did write about it, though at a slightly later date: Marcel Proust suffered poor health all his life and became an eager convert to the Théâtrophone, describing his experience in an evangelical letter to friends:

Are you subscribing to the Théâtrophone? ... From my bed I can be visited by the streams and birds of the Pastoral Symphony, which poor Beethoven himself could not enjoy any more directly than me because he was completely deaf. He consoled himself by attempting to reproduce the song of the birds which he could no longer hear. Even with my own lack of genius or talent, these are also pastoral symphonies which I create in my own way, whilst painting what I can no longer see.[16]

Proust died in 1922, sadly just too soon for radio, but his words stand alongside those of Edith Davidson in Lambeth Palace or Mr James in his grocery shop in bearing early testimony to a revolutionary new listening experience. At first glance, Proust seems simply to be suggesting that one sensory experience can be replaced by another. Interestingly, though, this occurs not when he listens to music played on that older technology, the gramophone player, but only with live, outside performances piped into his room. The Théâtrophone is so welcome in his confined lifestyle because it opens up inner chambers, extending or expanding the dimensions of the human mind itself. We do not all live our lives with the intense interiority of Proust; however, he was the first to voice a now familiar idea about radio, which can also be summed up by a distinguished contemporary British poet and playwright who recollected that in his Leeds boyhood 'the best plays I ever saw were on the radio'.[17] At a more mundane level, anyone who has depended on the radio while sleepless at night or laid up all day with a long illness or broken bone can relate to Proust's pleasure in the 'pleasure telephone'.

⸙

During the years after the death of the old queen and as the Edwardian era gathered pace, take-up of the Electrophone increased a little. In its launch year there had been just 50 subscribers, but this number would increase to over 1,000 by the end of the First World War and peak at a little over 2,000 in 1923, the year after the BBC was launched. At that point, a slightly desperate director of the company announced in a newspaper article that it would be a long time before broadcasting by wireless would reach the degree of perfection now offered by the Electrophone.[18] Just how wrong he was can be seen by a comparison of its take-up with the astonishingly wide and speedy spread of radio in its first seventeen years. Already in 1922, around 150,000 British people were able to listen to radio; a decade later this number had risen to around 20 million; when E.M Delafield declared in 1935 that 'the wireless installation is the background against which the whole design for living of the present-day family unrolls itself',[19] the figure was almost 28 million, and by 1939 it was 34 million people in a population of almost 48 million.

Once again, the story of a parlour song provides insight into the cultural penetration of wireless. In 1854 an American composer and lyricist who went under the name Alice Hawthorne[20] had brought out a song focusing on how joyless the hearth would become without a maternal presence:

> What is home without a mother?
> What are all the joys we meet,
> When her loving smile no longer
> Greets the coming, coming of our feet?
> The days seem long, the nights are drear,

This gathering of Electrophone listeners took place in the reception room of the company headquarters in London in 1908. They were listening to a live relay from a nearby theatre.

> And time rolls slowly on;
> And, oh! how few are childhood's pleasures,
> When her gentle care is gone.

'What Is Home Without a Mother?' was no seven-day wonder: copies of the sheet music of really popular songs could sell in their tens of millions, performed and performed again and enjoying currency for sixty or seventy years. Even people who did not have a piano or a parlour in which to play it would recognize titles and internalize their lyrics and themes. It was, in fact, just over seventy years later that the fledgling BBC chose to invoke 'What Is Home Without a Mother?', clearly working on the assumption that its twentieth-century public would immediately recognize the mid-nineteenth-century reference and apply it to the arrival of wireless in their homes. Halfway down

WHAT IS HOME WITHOUT A RADIO?
Mother (to nurse): "Let the little darling listen to the Children's Hour, and then, when he's had his supper, the Radio Dance Band can play him to sleep."

A cartoon from the *Radio Times* in September 1926 sketches the impact of broadcasting in a prosperous home in which even the nursery boasts an expensive wireless set.

page 2 of the edition of the *Radio Times* published on 3 September 1926 you'll find a small rectangular cartoon. An apparently careless sketch, on closer inspection it portrays the scene with brilliant economy: an elegant woman, wearing a cloche hat and cocoon coat with an extravagantly oversized fur collar, is just leaving the nursery on the arm of her equally debonair husband. At the last moment she turns her head with a final instruction for the dumpy nursemaid who stands humbly next to her toddler charge: 'Let the little darling listen to the Children's Hour', the mother suggests, 'and then, when he's had his supper, the Radio Dance Band can play him to sleep.' Below the image, in capitals, the title provocatively asks: 'WHAT IS HOME WITHOUT A RADIO?' It's a small cartoon, not a sociological study, but the suggestion is clear: less than five years after the arrival of the BBC a home was simply not a home without a radio.

For a more expansive image, just look at the iconic first ever Christmas edition of the *Radio Times*, which arrived on the news-stands a year after the birth of the BBC, in 1923 (see the opening to this chapter). A family is depicted by their glowing fireside in a modern living room with an art deco mirror and clock on the mantelpiece. The parents are sitting in elegant armchairs – mother in a long white dress is still channelling something of the Victorian 'Angel in the House'; father is in a soft tweed suit and tie, smoke curling from the pipe in his mouth. The three children cluster informally around them – one boy cross-legged on the floor, the other leaning over his father, and the girl, her dark hair cut in a fashionable bob, her long legs unencumbered by Victorian frills and furbelows, is perched on an upright chair with her head to one side like an alert bird. They are dressed in rosy reds and browns and their faces glow in the yellow light thrown by the fire, but none of them is focused on the hearth. All five are swivelled towards the elegant horn speaker of the wireless – their eyes glazed, they are transfixed as they listen. For reassurance – or perhaps ironically – the caption along the bottom of the cover is a quotation from another popular parlour song: 'Just a Song at Twilight'. A brilliantly eye-catching piece of promotional advertising, it is full of the promise of what radio can bring to family life. It encourages early-adopter householders to indulge their sense of self-satisfaction whilst urging others to invest in a technology which was now no longer quite so new or unfamiliar. A decade earlier, in 1912, the most famous maritime tragedy of modern history had brought understanding of the power of wireless to every fireside.

THE MARCONIGRAPH

WIRELESS·TELEGRAPHY·ILLUSTRATED·MONTH·BY·MONTH

APRIL 1911 PRICE 2d

TWO

READY OR NOT?

> You've set my valves a-throbbing,
> My head-piece in a whirl,
> So turn your ear piece to me, love
> My wondrous wireless girl
>
> Song published in first Christmas edition of the *Radio Times*, 1923[1]

As newspaper boys took to the streets on the morning of Tuesday 16 April 1912 there was still considerable confusion about just what had happened in the North Atlantic on Sunday night. It was agreed that RMS *Titanic*, an ocean-going liner of almost inconceivable luxury, had struck an iceberg some time before midnight and sunk in the small hours of the morning, but the fate of those on board was still very much in dispute. The reassuring if prolix *Daily Mail* headlines declared:

> TITANIC SUNK. NO LIVES LOST. COLLISION WITH AN ICEBERG. LARGEST SHIP IN THE WORLD. 2358 LIVES IN PERIL. RUSH OF LINERS TO THE RESCUE. ALL PASSENGERS TAKEN OFF

but the *Evening News* was soon leading with a more succinct – and, as it turned out, accurate – statement: TITANIC DISASTER: GREAT LOSS OF LIFE. Coverage of the tragedy rolled on and on over the following

The launch edition of *The Marconigraph* in April 1911 portrays wireless as an exciting new form of communication rather than broadcasting. The first significant journal dedicated to wireless, it rebranded itself in April 1913 as *The Wireless World*.

DAILY SKETCH.

No. 969.—TUESDAY, APRIL 16, 1912. THE PREMIER PICTURE PAPER. [Registered as a Newspaper.] ONE HALFPENNY.

DISASTER TO TITANIC ON HER MAIDEN VOYAGE.

The Titanic leaving Southampton on Wednesday last on her maiden voyage to New York. On Sunday night she struck an iceberg 280 miles south of Cape Race, Newfoundland, and in response to her wireless messages for assistance the Allan liner Virginian, her sister ship the Olympic, and the Baltic raced to her help.

Captain E. J. Smith in command of the Titanic, who was captain of the Olympic when she was in collision with the cruiser Hawke off Cowes. He has been a White Star commander for 25 years.—*Daily Sketch* Photographs.

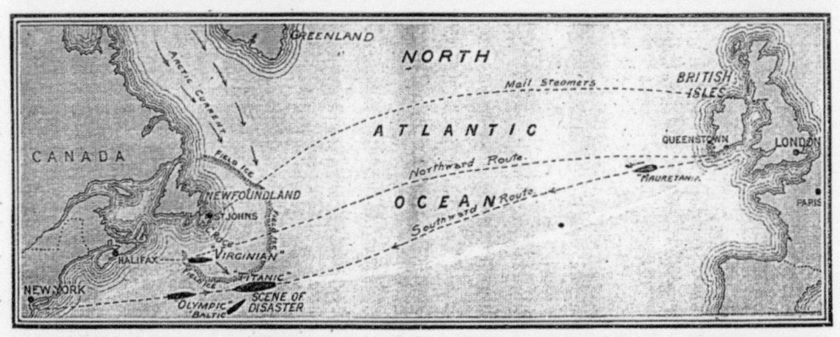

This map, drawn from the information contained in yesterday's wireless messages, shows the track of the great liner, and how the Virginian, which sailed on the homeward trip from Halifax on Saturday, was able to get to the Titanic first. The Olympic, which also sailed on Saturday, was able to steam towards the Titanic also when she picked up the wireless message. Many of the great liners traversing the ocean highway have been in communication with the White Star liner ever since her accident. The passengers were transhipped aboard the Parisian and Carpathia, which also got alongside yesterday.

The sinking of the *Titanic* on Monday 15 April consumed massive quantities of newsprint. This front page of British tabloid the *Daily Sketch* was published the day after the event.

days and weeks, but there was one matter upon which journalists, officials and politicians were in total agreement: how many more lives would have been lost had it not been for Guglielmo Marconi?

In his teen years Marconi had become obsessed by whether radio waves, a newly discovered type of radiation which had the same nature as light but a longer wavelength, could be used to transmit signals. In 1896 he had travelled from Bologna to London in the company of his Irish mother. He was just 21, but he was a young man in a hurry, and, in that same year, had been granted British Patent number 12039, 'Improvements in transmitting electrical impulses and signals and in apparatus therefor'.

Something of Marconi's personality, drive and ambition is still evident in a swordstick he enjoyed wearing at ceremonial occasions, of which there were to be many as his fame grew. Draw out the slender steel blade from its cane casing and you can still read the warning inscribed in archaic capitals: NON TI FIDAR DI ME SE IL COR TI MANCA.[2] These words come from a Tarot deck, hugely popular in Italy, and translate as 'Trust not in me if your heart be faint.' Their essential meaning is 'Don't handle dangerous weapons if your heart isn't up to it', but they refer, perhaps, also to Marconi's new invention – wireless communication – with its extraordinary but as yet uncertain potential.

Drive and ambition, though, are not the same as vision, and for a long time Marconi failed to recognize where his invention was leading. Ideas about the transmission of words and music were simply not in his mind, nor any concept of how the home might be transformed. In fact, the word 'broadcasting' was not yet in use except in its specialist agricultural sense of scattering seeds across a field, and it was still so unfamiliar in the early years of the BBC that there were high-level discussions about whether the past tense of the verb was 'broadcast' or 'broadcasted'. All this was in the (near) future. Marconi's focus was entirely on commercial potential and on the naval use of wireless telegraphy to transmit dot–dash Morse signals from ship to shore or ship to ship. Less than a year after obtaining his patent, he sent the first ever wireless communication over open sea. The message travelled across the Bristol Channel from Flat Holm Island to Lavernock Point near Cardiff, a distance of 6 kilometres. It read 'Are you ready'. Though it contained no punctuation, the question was soon being answered in the affirmative.

First-class passengers on the *Titanic* might be looking forward to enjoying its Turkish baths and Parisian-style café, but one of the first things many did when they embarked was to take advantage of the new wireless technology service offered on board and send messages to their friends back on shore. Four days out of Southampton, she struck the iceberg. Unfortunately, there were as yet no established rules that clear wireless channels must be kept for emergency communications: the *Titanic*'s operators were transmitting over the same frequency as other ships and, because the channels were also jammed with these personal communications, several ice warnings from other vessels were either ignored or missed. To make matters worse, wireless operators worked only day shifts and then closed down for the night, so it was sheer good fortune that an operator on another ship, the *Carpathia*, happened still to be awake and eventually received the *Titanic*'s distress call. In tragic contradiction of the *Daily Mail*'s optimistic headline, 1,500 of the passengers and crew were lost. But as news filtered through that 704 had been saved, the Postmaster General summed up the opinion of the press and the nation when he declared that without wireless telegraphy there would probably have been no survivors at all. Relieved not to have been on board, newspaper readers now understood that 'those who [had] been saved, [had] been saved through one man, Mr Marconi ...and his marvellous invention'.[3] The catastrophe hugely boosted public awareness of the wonders of wireless: within a year, even children in school playgrounds were singing songs about it. Although nothing equalled the sinking of the *Titanic* for all-round drama, just eighteen months later, in the autumn of 1913, news of another maritime disaster began arriving. The *Volturno* was just an average-sized ocean-going liner and her passengers were largely bedraggled Russian and Eastern European Jewish refugees, desperately fleeing persecution and hoping to make new lives for themselves in America. The ship set sail from Rotterdam, but halfway through its two-week journey she was rocked by a massive explosion. The dawn seas were stormy so nearby ships took time to come to the rescue of the *Volturno*, but the ratio of survivors was impressively high: of the 657 passengers and crew, 521 were saved. This too made a deep impression

"S. O. S."

PUNCH (*to Mr. MARCONI*). "MANY HEARTS BLESS YOU TO-DAY, SIR. THE WORLD'S DEBT TO YOU GROWS FAST."

This *Punch* cartoon from October 1913 shows how the *Volturno* disaster, like the sinking of the *Titanic* in the previous year, provided a publicity coup for Marconi and his company.

on the public. A famous cartoon by Leonard Raven-Hill depicts Mr Punch standing, hat in hand, before Marconi, who is seated in front of a panel of technical apparatus. He holds a set of headphones in his hand and his workmanlike desk is scattered with wireless components. Beneath the image are written the letters 'S.O.S.' and then the words addressed by Mr Punch to Marconi: 'Many hearts bless you to-day, sir. The world's debt to you grows fast.'[4]

In reality, Marconi's own days of sitting at a humble corner desk experimenting with radio components were by now a thing of the past: he had, after all, been the joint winner of the 1909 Nobel Prize in Physics. But this is precisely what he had once done, and there was now an eager cohort of adolescent British boys and men for whom wireless communications had become an all-absorbing new hobby. One of them, John Clarricoats, later fondly recalled the kinds of things they got up to:

> They were the young men who built small electrical motors ... laid telegraph wires to the homes of their friends across the street, experimented with telephones and made up electric batteries. They had been fascinated by Marconi's successes, both in his native Italy and in England. They read all they could about his work, talked about the results and, in their own way, tried to repeat his experiments in their own homes.[5]

Another, Arthur F. Carter from Buckinghamshire, improvised his first transmitter in 1912 with the use of an old motor-car ignition coil, a pair of brass doorknobs, a copper helix, some photographic plates and a sheet of tin foil.[6] Most of the names of these so-called hobbyists have now been forgotten. Not so Peter Eckersley. He was later to become the BBC's first chief engineer, but he was bitten by the wireless bug as a teenager. Returning from boarding school for the holidays, he discovered that his brother was already a step ahead:

> I found our playroom filled with lovely and exciting instruments. There were Induction Coils to make fat sparks, Leyden jars, long black rods of ebonite wound with green silk-covered wire, X-ray tubes and galvanometers. The things, their touch and shape, gave me a sensual pleasure and made me want to understand what they were for...[7]

Eckersley was a pupil at Bedales, an exclusive boarding school, but many wireless hobbyists came from more modest backgrounds and experimented in spartan bedrooms or draughty garden sheds. At the lectures and meetings of the wireless clubs and societies which were being formed across the country, they were united in their passion for this new technology. Having constructed their sets, though, what was the point of it all? The term these boys and men used was 'listening needs no technical knowledge. The person who intends going in for experimenting can start straight away with only a small amount of technical knowledge which he can pick up from the two books mentioned above.

a kind of electric lamp known as a valve. This valve has four prongs, which fit into a special valve holder, and it is necessary to provide two kinds of batteries to work a valve set. One battery is usually an accumulator which

At a wireless outing. Wireless societies do a great deal of useful work. *You* should join.

Valve and Crystal

As regards the apparatus used in receiving, there are, in general, two classes of wireless receivers. We, first of all, have the crystal set and we also have valve sets. Some valve sets also use crystal detectors.

A crystal detector merely consists, in its simplest form, of a piece of mineral on which presses a light wire spring, generally known as a "cat's whisker." This crystal detector is fitted in a box which also contains a coil of wire called the inductance. Means are provided for tuning the crystal receiver to the particular broadcasting station which you desire to receive. On a certain adjustment of the set one station will be received, while with another adjust-

can be recharged with electricity at suitable intervals; the other battery is called a high-tension battery and gives about 60 to 100 volts. These high-tension batteries will last several months and are then useless and a new one requires to be bought.

Books You Need

It is really impossible to enter into a detailed technical explanation of wireless in a magazine of this kind. The best thing you can do, unquestionably, is to buy the first two books of the Radio Press series called, respectively, *Wireless for All* (price 9d.) and *Simplified Wireless* (price 1s.). These two books, the second one of which is a sequel to the first, are written by Mr. John Scott-

This group, almost entirely male, has set up a large portable aerial for outdoor listening. The photograph was published in a magazine for fellow wireless constructors in November 1924.

in'. It's a phrase which lasted well into the first two decades of radio, and by then it simply meant 'to listen' as we listen to radio today, but back then it was all about tuning and tuning again their receiving apparatus until they could pick up Morse signals from a transmitter somewhere in the country or abroad. One early enthusiast, J.E. Nickless of Wanstead in London, proudly recalled sitting for eight hours and hearing just one time signal, transmitted from the Eiffel Tower in Paris.[8] This sort of outcome satisfied their particular passion but it was essentially a form of eavesdropping. Even those who were willing to predict a more popular future for wireless assumed that this would involve Morse-code messages being 'translated' into speech and then shared with public audiences. When radio broadcasting was introduced less than a decade later, its most significant characteristic was that it was transmitted into individual homes, but this was far from the vision of Alan A. Campbell Swinton, the inaugural president of the London Wireless Society, in January 1914:

> Wireless seems by its nature to be suited for the unusual distribution of intelligence, weather reports, time signals or speeches. Indeed, with a little imagination, one could picture in the not-too-distant future wireless receiving stations specially set up in connection with halls, resembling picture palaces, where people would be able to go and hear viva voce all the prominent speakers of the day, although they might be speaking hundreds of miles away.[9]

Even the most committed hobbyists needed some downtime. In April 1911 the first magazine specifically dedicated to wireless, *The Marconigraph*, appeared on the news-stands. Two years later it was rebranded with the more mainstream title *The Wireless World*.[10] As well as providing technical information, each number featured a short story. Nobody could argue that they merit much literary attention, but these stories reveal how the concept and even the specialist vocabulary of wireless were beginning to permeate the public imagination, making its way into common usage.

The launch edition contained a story titled 'The Pleasure Pilgrims'.[11] When the juvenile lead, a young barrister called Harry Stenhouse, inherits a fortune, he books himself a stateroom on the bridge deck of a luxury ship about to depart on a world cruise. Exactly a year after the sinking of the *Titanic*, any reader would have nodded with approval at the description of the ship's watertight bulkheads and wireless telegraphy cabin. In fact, since *The Wireless World* had a keen readership among boys and young men hoping to train as wireless operators themselves, the true protagonist of the story is the oracular young Marconi operator, Tommy. The action begins on the first morning out of Southampton, when Harry encounters the lovely Nessie Lynwood, travelling with her widowed mother. He is soon attempting to woo her, but after various failed advances he retires to the Marconi cabin:

> 'Now Tommy,' said Harry, 'You have been twice across the Atlantic, and have had a lot of experience of affairs. What would you do if you could not get any reply to your signals from a young lady because she was

too busy chaperoning her mother?' 'Have you tried the S.O.S. signal?' Tommy asked.

'Oh, yes. But apparently she knows neither Morse nor remorse.'

'Then you must get the mother married and settled,' replied Tommy.

'You are a bright boy,' commented Harry, 'I wish you would marry her yourself.'

'Personally,' responded Tommy, 'I do not think that men of science ought to marry. At any rate, I shall hardly have time for it. In another ten years, Marconi and all those chaps will be thinking of retiring, and there will be a chance for fellows like me...'

Suffice to say that Harry soon notices that Mrs Lynwood is falling for and at risk of marrying a newly embarked passenger who turns out to be a bigamous swindler and may also harbour Fenian sympathies. It is of course Tommy who alerts the police, using his skill and considerable luck to send what he calls a 'freak message' that travels hundreds of miles beyond its expected distance. At the close of the story, as Harry kisses Nessie, now his fiancée, the author comments on their encounter in innuendo-laden language of a kind which was about to become commonplace: 'although a powerful transmitter was close to a sensitive receiver, it was apparently found necessary to repeat the message an indefinite number of times.'

Perhaps recognizing that the creative well of this anonymous author was already running dry, the editors now commissioned a long serial story by one Bernard C. White, which was to run through the rest of 1913 and into 1914. The hero of 'A Pawn in the Game'[12] is Charles Summers, the son of a country parson, who is lucky enough to have a garden shed in which he does more than just tinker – he actually invents and builds a wireless-controlled airship. Charles's fiancée, Gwen Thrale, is 'a bright, intelligent, original girl, the idolised daughter of the squire', but also, after a few years spent in London, 'secretly the member of a Fabian Society'. She absorbs much of Charles's work, including how to operate the airship, but this 'mass of soft fluffiness' ultimately becomes the eponymous pawn in the game. Through her dangerous socialist connections, the blueprint for the airship falls into the hands of two dastardly foreigners who have inveigled their way into the village. Gwen does get to play a surprisingly active part in foiling them, but ultimatelys it turns out 'everything was too big and her mind could only grasp little things'. In the closing scenes she acknowledges the damage she has caused: 'I can see what a silly little fool I was.'

Even more sinister, as the radio historian David Hendy has brilliantly discussed,[13] was the serial's rabid xenophobia. So-called 'Invasion Literature' had been around since 1871, whipping up and feeding national anxieties about the risk of the country being overrun by foreign powers. Now that Marconi was a household name, fears about the power of wireless – in the wrong sort of British hands or pretty much any foreign hands – were creating a new literary subgenre. Dark clouds were already beginning to overshadow the sunny idyll of Charles Summers and Gwen Thrale's courtship. Three months after the triumphant conclusion of the story, on 28 July 1914, war was declared.

With wireless about to play a key military role in the conflict, all permission to experiment was abruptly withdrawn in accordance with the Defence of the Realm Act. As soon as the war ended, though, the speed with which it took off as a domestic medium – and a very British one – was truly astonishing.

It's not hard to imagine the excitement of boys who had grown to manhood during the war years as well as conscripted men who had been trained as Morse-code wireless operators: back in their sheds and bedrooms, they now found themselves able to pick up broadcasts of a totally new variety. Transmitted from two lofty aerial masts erected at Marconi's factory in Chelmsford in Essex, listeners with the right equipment could now tune in to experimental broadcasts of speech and music. Various members of Marconi's office staff were roped in to perform these – clearly well beyond their job descriptions but surely with an air of excitement: two assistants on the cornet and the oboe, a research engineer on a one-string fiddle, a senior engineer on the piano. The first British radio soprano – and, as far as I know, the first female voice on British radio – was another member of staff, though Miss W. Sayer was rather on the wrong side of history when she dismissed wireless as a 'punch and judy show'.[14] Listeners-in might also catch the nightly ritual of Marconi's engineer, W.T. Ditcham, reciting the names of the main British railway lines and London stations,[15] as well as two daily half-hour programmes consisting of news, further ad hoc musical performances and gramophone records. They themselves were proud to play a role: those who picked up broadcasts were urged to send in postcards reporting the technical reception on their home equipment. Voices and music were now arriving wirelessly into the homes of these pioneers, tiny in number but growing fast.

The press at once took an interest: 'The imagination is attracted to the fantastic possibilities of the wireless "phone"', wrote a reporter from the *Liverpool Echo*,[16] when he was sent to interview Mr Southwood, a radio hobbyist in New Brighton, about 'this most modern of scientific marvels' and to listen in to the programmes he was able to pick up. The story was run under the headline 'Bridging Space by Aerial Waves – Liverpool Man and Air 'Phone Experiments – Sung in Chelmsford, Heard Here'. The journalist then described how, by delicate adjustment of his wireless apparatus, Mr Southwood had enabled him to hear 'Land of Hope and Glory' sung by a quartet at Chelmsford. He was savvy enough immediately to point out the obvious attraction of 'plain speaking over a dot–dash method of communication'. Finally, under the subheading 'Talking Teapot', he reminded readers that it was not even necessary to have a wireless mast and aerial but that any mass of metal, including a spring mattress or a kitchen utensil, could be used to capture the incoming airwaves.

Stories such as this one, printed on page 5 of a regional newspaper, were whetting the public appetite for radio, but then Lord Northcliffe, the hugely powerful proprietor of the *Daily Mail*, who had a love–hate relationship with the medium, came up with a spectacular publicity stunt: his paper would sponsor the first ever broadcast of a live professional concert by the most popular opera singer in the country.

Nellie Melba was now almost 60 and had recently been made a Dame, so perhaps it wasn't surprising that, at first, she haughtily declined the invitation: 'My voice is not a subject for experimentation.'[17] But then Northcliffe offered her £1,000 (about £45,000 today), which, even by her standards, must have seemed easy money for a twenty-minute recital. On Tuesday 15 June 1920 she travelled down to Chelmsford where she performed what she was later to describe (she was, retrospectively at least, generally on the right side of history) as 'the most wonderful experience of my career'. Before long, this simply became known as 'The Melba Broadcast' and was heard by listeners across Europe and as far away as Newfoundland. Now Dame Nellie was all over the national press, her pale face under its wide picture hat, her substantial bosom and tiny square handbag dangling from her hand on a delicate chain. She stood, for the broadcast, in front of an unfamiliar contraption called a microphone, which looked like a kitchen measuring funnel tipped on its side but had in fact been improvised from a telephone mouthpiece and the wooden panels of a cigar box. The number of people who actually heard the concert was small, but the whole country knew about it. The *Liverpool Echo* reporter told his readers how it began with 'Madame's' famous long silvery trill – she later described this as her 'hello to the world' – followed by 'Home! Sweet Home!' in English, Maurice Bemberg's 'Nymphs et Sylvains' in French, Mimi's aria 'Addio' from Puccini's *La Bohème*, sung in Italian, before concluding, of course, with a rendition of 'God Save the King'.[18]

Having set the scene with a description of some 'atmospherics' caused by local cloud cover, the reporter then launched into a vivid description of what he had heard, sharing the experience with the majority of readers (perhaps including my grandparents and father), who would not have heard it themselves: 'Suddenly, into the comparative quiet broke a voice, matter-of-fact and ordinary, announcing that the great lady would sing, and once she launched into her short programme, it was as though we were in a large concert-hall with Melba at the far end.' The *Daily Mail* announced the event a triumph, an unprecedented coming together of art, romance and science. Other papers concurred, largely writing from the broadcasters' perspective – 'Chelmsford to the World!' was the sort of bathetic tone – but the Liverpool reporter was more pragmatic in understanding what

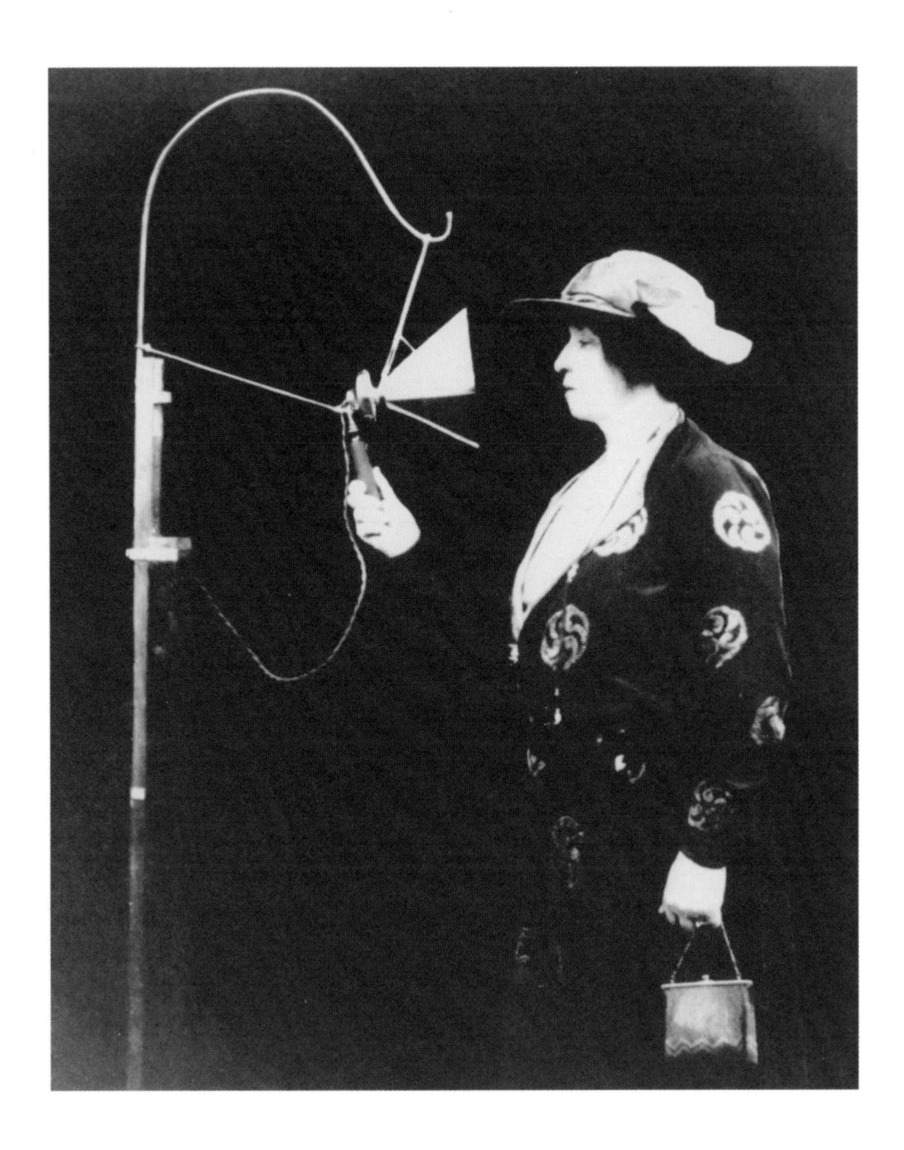

History sadly does not relate what Dame Nellie Melba kept in her tiny handbag, but the look on her face suggests the novelty of wireless when she made her famous broadcast from Chelmsford, Essex, on 15 June 1920.

wireless would really mean to ordinary households when he suggested that the experience of hearing Melba sing in a live concert hall might now be brought to listeners sitting by their firesides in their own front parlours.

❧

This was only the start. Eighteen months later, on St Valentine's Day 1922, the hobbyists who slipped on headphones and tuned their apparatus picked up a truly landmark broadcast: they had heard amateur one-string fiddle-players, they had enjoyed Dame Nellie's premiere, but this was what we would recognize today as an actual programme. From a wartime prefabricated hut in Writtle, a village close to Chelmsford, and using the call signal 2MT, a new team embarked upon their appointed mission of transmitting a weekly experimental half-hour programme for all wireless 'amateurs'. At their helm was Peter Eckersley. Since the war, he had acquired the rank of captain, but the tone was astonishingly subversive. Rejecting the official plan, which was mostly to play gramophone records sent down from head office in London, they soon began improvising. One of the team later recalled their freewheeling approach:

> Some time on Tuesday afternoon the piano would be trundled into the hut, and we would receive a bunch of records – most of which were usually rejected as too highbrow! Programme planning was done at the 'Cock and Bell' up the road, about half an hour beforehand. We had artistic ambitions – for example, we put on Cyrano de Bergerac, the first play ever to be broadcast in this country. I well remember our sitting round a table in my digs, reading our scripts with spoons held to our mouths in simulation of the hand microphones we had to use. There were more players than microphones, so much of the rehearsal consisted of practising the passing of a spoon from one to another without dropping it. But our star was Eckersley. He'd go up to the microphone, and apparently without effort, be spontaneously funny for ten minutes at a time. He talked to our listeners as if he'd lived next door to them for years, and they loved it.[19]

Sadly, no audio survives, but listeners in their bedrooms and sheds must quickly have recognized that this was something entirely new. Eckersley finished each show with a theme song, his high tenor voice accompanied by raucous piano playing:

> Dearest, the concert's ended, sad wails the heterodyne,
> You must switch off your valves, I must soon switch off mine.
> Write back and say you heard me, your 'hook-up' and where and how,
> Quick! For the engine's failing, goodbye you old low-brow.[20]

Wireless listening as after-dinner entertainment in the home of the adventurous Miss Florence Tyzack Parbury in May 1922, when it was still unusual to see three women at the centre of a photograph on the theme of broadcasting.

This request for responses was actually the point of the exercise: after the first show, more than fifty listeners wrote in with their reactions and the postcards continued to arrive week after week. Even at its height, there were perhaps only 100,000 listeners to Writtle[21] – a tiny figure, but it was becoming clear that there was more than a specialist audience out there.

On 20 May, for example, *The Graphic*, a weekly illustrated newspaper, ran a story recommending a wonderful 'diversion from the orthodox after-dinner programme – listen by means of a simple arrangement of coils of wire to music, drama, revues, debate, all part of this great development in the art of annihilating space'. The accompanying photographs shows the socialite, author, musician, painter, traveller and aviator Florence Tyzack Parbury in full evening dress, surrounded by a group of men, some smoking after-dinner cigars, as they listen in to a boxing match. Few people any longer know about *The Graphic*, or (sadly) the truly intrepid Miss Tyzack Parbury, but the name of the broadcaster was to become rather more familiar. The success of 2MT had been so extensive, so unexpected, that Marconi had opened a sister station in London, and the boxing match between Georges Carpentier and Ted 'Kid' Lewis had been part of its first evening's broadcasting. The official launch of this new station was later in the year, when, at just after half past five in the evening of 14 November 1922, listeners heard a considerably more sober voice – the precise consonants and rounded vowels of Marconi's former press officer Arthur Burrows – enunciating the new call signal: '2LO, Marconi House London calling'. This was followed by a brief news bulletin, read once fast and once at dictation speed so that listeners could take notes. Soon 2LO would simply become known by its more familiar name, the British Broadcasting Company (later Corporation), or BBC, under the command of a young Scot named John Reith.

Florence Parbury was a woman of great wealth, but even the vast majority of people, for whom a set was still an unimaginable luxury, couldn't fail to notice that images of wireless were suddenly everywhere. The manufacturers soon realized that one photogenic gimmick could move masses of units if picked up by a newspaper or magazine: it was apparently now possible to buy a set hidden inside a top hat or a Christmas cracker; an enterprising society hairdresser in London's West End dreamed up a 'listening-in coiffure' – two rather unbecoming plaited side buns which slightly resembled headphones.[22] If you weren't a reader, you could listen to the great stream of wireless-themed novelty songs. One of the first of these, by Harry Pease, was sung by G.H. Elliot, a music-hall favourite who had long been popular as a minstrel performer of numbers such as his racist yodelling song 'Lily of Laguna'. His days of blacking up (shockingly and astonishingly, even on the radio) were, sadly, far from over, but he didn't hesitate to leap onto the bandwagon. 'Virginia' had once featured in his songs as the American state which had shamefully clung to slave ownership, but it was now converted into the name of a modern young woman whose beau was calling out to her, imploring her to pick up his radio call:

Listen in, Virginia,
To someone who is feeling blue;
Listen in, Virginia,
'Cause every word is meant for you;
Anybody, everybody, 'way down there,
I don't care who they be;
Somethin's doing, somethin's brewin' in the air
So please tune up on me.[23]

At a slightly more literary level and with a fashionable dash of syncopation (particularly in the chorus – at the start of this chapter), Christopher Marlowe's Passionate Shepherd could now be heard addressing his Elizabethan lady in the language and technology of the twentieth century:

O come with me a-listening, love,
O heart of mine!
Let's float amidst the ether, love,
With arms a-twine.
With souls a-tune,
O rose of June,
Our love like crystal set
Give ear, my peach
To my wireless speech
And listen with me, pet.[24]

There was also a rush of new celebrity endorsements, each appealing to different tastes and pockets: the famous prima donna Madame Tetrazzini was shown smiling broadly in a splendid feathered hat as she tested out a top-end set; *The Sketch* featured Miss Dorothy Dickson, 'now delighting London as ... the heroine of The Cabaret-Girl at the Winter Garden', who was apparently now the possessor of a wireless set.[25] She was 'badly bitten with the new craze and spends a good deal of time listening-in'; the light opera star Miss Evelyn Laye allowed herself to be photographed listening-in while still in costume as the eponymous Merry Widow.[26] There was also (a personal favourite of mine) Miss Gladys Walton, the film star, 'listening to wireless gossip by means of her garter'[27] (in fact a new piece of technical gear from America intended for airmen who needed hands-free sets, but why let facts get in the way of displaying a girl's legs?); and the famous

Oct. 17, 1923 *The Sketch* 131

Tuning-In: The Merry Widow and Her Wireless.

SONIA LISTENS TO THE CALLING OF 2 L O: MISS EVELYN LAYE.

Miss Evelyn Laye, the charming young light-opera star, who is having such a big success in the title-rôle of the revived "Merry Widow," at Daly's, is—like everyone else—interested in broadcasting, and is shown in our photograph, attired in a Merry Widow dress, enjoying a few moments' "listening-in." Miss Laye, who made her first appearance on the stage at the age of three, has been seen in a number of successful London productions, but the Merry Widow represents her first important rôle in a light opera—and is not in the least likely to be her last, though, judging from the success of the revival, it will be some time before she is seen in another piece!

Photograph by Foulsham and Banfield, Ltd.

By October 1923 magazines and newspapers were crammed with images – the more glamorous or quirky the better – of wireless listening, such as this one from *The Sketch*.

dancer Miss Phyllis Bedells listening in with a friend while relaxing in the elegant upright coachwork of a Daimler Landaulette.[28] It was even possible to listen in while on the move in a considerably less upmarket vehicle: *The Sketch* shared a photograph and described a 10-mile pleasure trip between Bournemouth and Wimborne on a charabanc or 'charry' which was equipped with headphones for every passenger.[29]

Interestingly, more than one newspaper featured a story about a death which occurred during or after a broadcast. The newly widowed Mrs Cuthbert Aubrey Judland, for example, gave this poignant account at the inquest for her late husband, a 52-year-old export agent's clerk from North Kensington in London: 'We were in bed listening-in on the wireless, and when the programme had finished, he took off his headphones with the remark "That has been very nice". He then collapsed and died.'[30] The shock of loss must have been terrible, but there's a suggestion here that wireless listening was already becoming a comfortable addition to home life.

Newsagents began stocking a different style of wireless magazines which were devoid of jargon. The cover of the first number of *Popular Wireless* showed a fashionably dressed young man and woman (there is not a touch of Gwen Thrale's supposed 'mass of soft fluffiness' about the woman) – perhaps siblings – with an older woman who is tatting. All three wear headphones and are listening to a broadcast. Above the masthead are the words 'All about Wireless' but also 'A New Paper for ALL', and below it a reassuring promise: 'The World's Latest Hobby Fully Explained'.[31] It all seemed perfectly unobjectionable, but one already experienced group was far from happy. These were the hobbyists, and they were actually pretty indignant when they realized that this swiftly growing number of ordinary listeners, who had no technical knowledge and – worse still – were shamelessly indifferent to the technical aspects of wireless, might interfere with their own experimenting. It was clear that such broadcasts challenged their hegemony and reduced the bandwidth available to them (a serious problem in the pre-digital age). Understandably, this all felt rather unfair when they had volunteered their services during the serious

The launch edition of *Popular Wireless* in June 1922. The wording, underlining and image itself all suggest the expanding interest in and marketing of broadcasting.

development stages of wireless. They also suggested that when issuing licences (a key part of the funding model for the new technology) the Postmaster General should make a distinction between valuable experimenters such as themselves and members of the public who were buying ready-made sets merely to listen to broadcasts for the sake of amusement.

HOME
RADIO
6D.
A clear and non-technical explanation of
WIRELESS
With full instructions for beginners.
How to make, fit up, and use at home.

In spite of the suggestion of glamour emitting from the expensive-looking set, this undated early magazine was cheaply produced and clearly intended for less prosperous would-be purchasers.

A wireless set just for amusement? In spite of their outrage, it was already clear only a decade after the sinking of the *Titanic* and the *Volturno* that the function of wireless had undergone a transformation. More than that: although the idea of buying a receiving set for domestic use was still a novelty, the newspapers were eager converts. Precisely a year after the first of Peter Eckersley's experimental broadcasts from Writtle, the *Aberdeen Press and Journal* was boldly declaring: 'whether you can spend thirty shillings or thirty pounds, you can have a wireless set.' There was no end to the extravagance of the more expensive sets, but cheaper ones were also accessible. Would listeners from ordinary households decide to fork out, though, and – most importantly – would they find the outlay was enhancing their home lives?

THREE

WHAT ARE THE WILD WAVES SAYING?

> All you could hear was the sea, you know, like the sound of waves – but oh, there was such a hullabaloo if you could hear one voice, just one voice.
>
> Housewife, Warrington[1]

Bunty must have been brimming with excitement as she and her father set off on Wednesday morning, 23 April 1924.[2] Everyone had been reading and talking about the British Empire Exhibition. They knew that Queen Mary's own dolls' house would be on display, with its miniature working piano and flushing lavatories; that there would be a full-size sculpture of the Prince of Wales, carved out of Canadian butter and displayed in a refrigerated glass case; and that visitors could even eat a meal in the Indian pavilion, which had been built to resemble the Taj Mahal.

The idea for an exhibition to showcase the natural resources and industry of Britain and her empire had been under discussion since the start of the century and postponed more than once. Now, on St George's Day, its gates were at last being thrown open. Even King George V must have felt some excitement as he travelled to the venue with Queen Mary at his side. Of course, he was always opening big national events, but today was different. This time the king was not

Of the seven girls in this iconic 1923 image entitled 'Listening to *The Children's Hour*', two are wearing headphones, two are using single-cup extensions, two are trying to catch any sound and just one is entirely left out – the equivalent of today's 'digitally excluded'.

only going to be speaking to 120,000 people in the showground at Wembley, then a suburb north-west of London, but to his people across the nation. In Croydon, south of the Thames, Bunty and her father joined a crowd which soon became so large that the police had to stop the traffic.

The king's speech was not only the first live radio broadcast by a British monarch; it also happens to be the earliest surviving recording of any British radio broadcast. In classic British style, the king acknowledged how progress in building had been hampered by exceptionally unfavourable weather, before going on to speak of his belief in the potential of the Empire. As he declared the event open, his words were greeted by roars from the crowd, while 'the whole audible pageant', as it was described by a *Daily Mail* journalist, 'was heard across the country':

> the measured and clear voice of the King [was] conveyed by wireless to a vast audience far out beyond the shadows of Wembley. Those who heard realised perhaps for the first time that wireless has not so much extended the power of one sense, hearing, as presented a new faculty to man.

The broadcaster of this event was the BBC, which had by now been in existence for eighteen months, but the *Daily Mail* had cleverly capitalized on it by teaming up with Marconi's company. Head office in Chelmsford had cancelled all Easter leave for engineers and technical assistants whose services might be called upon in London,[3] but this was a nationwide affair, with loudspeakers set up across the country. Some were in halls, but most were in the open air, so as to accommodate the anticipated crowds. In Inverness, for example, the loudspeaker was attached to a German gun, captured in the last war, up on Castle Hill, where three thousand shared this national experience:

> An extraordinary feature was the 'realness' that characterised the proceedings. Crowds all over the British Isles stood bareheaded while the massed bands played the National Anthem at Wembley, knelt while the Bishop of London delivered his prayer, and joined in while the Lord's Prayer was recited in the Stadium. People had anticipated something different from what they had ever seen or heard before, but never, in their most sanguine moments, did they realise that they would hear everything

> so clearly and so realistically. As the programme went on they could almost reconstruct, in imagination, the scene being enacted at Wembley.[4]

The reporting may be a touch overawed (the *Mail* had invested a good deal of money) but this was indeed a memorable event. In the following days, brief reports, largely though not universally positive, were filed from loudspeaker stations across the country: Aberdeen ('Several thousands present. A huge success'), Brighton ('10,000 people heard the King's speech in front of Theatre Royal, and road was closed to traffic by police') and Colchester ('Hearing indifferent: 2000') to Stoke on Trent ('Great crowd from pottery factories listened to speeches in the rain'), Thame ('Very large crowd. Reception enthusiastic') and Warwick ('Very successful. 600').[5]

The *Daily Mail* was soon receiving letters of thanks. We know about Bunty because she supposedly penned one of these: 'My Daddy took me to the place in Croydon to hear ... our King open the exhibition. We both heard beautifully and we shouted out "Hooray".' Meanwhile, Peter Eckersley, now the BBC's chief engineer, announced on air that telegrams had been received from John O'Groats to Land's End reporting that the speeches and the whole ceremony were distinctly heard.

THEN—Queen Victoria and Prince Consort at the 1851 Exhibition

On May 1, 1851, Sir Joseph Paxton's "Crystal Palace" was opened by Queen Victoria in Hyde Park. The above picture (an early photograph) taken seventy-three years ago, shows the Queen and her Consort on the platform at the opening ceremony. The Queen's voice on the platform was heard by some few hundred spectators, whereas—

And NOW—Listening to the King's Speech in Edinburgh

—At the opening of the British Empire Exhibition, not only was the King's voice heard throughout the vast audience of 100,000 in the Stadium, but, thanks to the employment of modern science, was heard throughout the length and breadth of Great Britain, and, if report be true, by at least one listener-in in America

These two images from the illustrated newspaper *The Sphere* contrast the Victorian era with 1924, showing how the public could share far more closely in royal and national events through wireless broadcasting.

There is, of course, just a possibility that little Bunty and her gratifyingly specific message were inventions of a journalist writing to a deadline, but there's no doubt at all about the historic nature and impact of the event. As many as a million British people may have borne witness to this broadcast – the first time they heard the king's voice, but also, far more importantly, their first personal experience of listening to wireless. The vast majority were in these public crowds. In the following week's edition of *The Sphere*, the picture editor cleverly juxtaposed two images from 'then' and 'now'. The double caption contrasted the mere few hundred who had heard the voice of Queen Victoria at the opening of the Crystal Palace back in 1851 with the 100,000 who had been in attendance at the Empire Exhibition, but it also drew attention to the vast number of listeners across the country, such as those in the second photo, who were gathered, far from Wembley, around a loudspeaker set up on a bandstand in Edinburgh.[6]

Some, though, were listening in their homes, and it is these 'early adopters', as we would now call them, who were the true trailblazers and forebears of today's radio listeners.

Almost all of them listened on the most popular form of home receiver – the crystal set, named after its key component, a mineral crystal detector which had to be connected with a tiny coil of copper wire, known as a cat's whisker, in order to tune the gadget and pick up the signal. These had no amplification, so headphones – or 'earphones' or 'telephones' as they were also called – were an essential accessory. Some constructed more ambitious sets from kits, which would arrive in the post, complete with blueprints. In 1988 Shaun Moores had the brilliant foresight to conduct a series of interviews with elderly people in his home town of Warrington, capturing their memories of early wireless: 'Uncle Bill made our first set from a kit', remembered one interviewee, 'Oh, he had diagrams and goodness-knows-what for the kits. He used to get the components and piece them together. Uncle Bill was a bit of a one for hobbies.'[7] For those with even smaller purchasing power or only pocket-money savings (my father among them), there were other options. Moores includes this moving memory from a devoted wife:

> When it was his birthday, or when Christmas came, I used to give him parts for his wireless, d'you see. I'd put fourpence away every week to save up to get the bits he was after. All the family – not just me – bought him these different parts for it and he built it up himself.[8]

A mahogany two-valve Marconiphone Type V2A from around 1923. Compact and handsome, it still resembled a scientific gadget rather than a piece of furniture.

Crystal sets could be bought for between £3 and £5. Requiring no batteries and costing nothing to run apart from the annual fee of 10 shillings for a 'receiving licence' (which remained unchanged throughout the period), these were the ultra-economy option. An agricultural labourer might bring home only 25 shillings a week, a train driver 50 shillings, and a miner £4, so such sets were the only ones within their purchasing power, particularly with constant downward pressure on wages. At the other end of the scale, newspapers and magazines were soon carrying advertisements for ready-made wireless sets in the £20 to £60 bracket, and there were plenty of even fancier ones on show. Colourful images appeared in periodicals such as the *Illustrated London News* of well-dressed families clustered round the newcomer in opulent drawing rooms: in one, a mother with her children and their friends are taking part in another new craze, a 'wireless party'. They look on as the father demonstrates the new device, complete with a large, table-top frame aerial and amplifying horn.

In real-life homes, where the furnishings of the Victorian parlour still lingered, such a crude scientific object was often considered unsightly. Designs for this new era of wireless were all about disguise – tidying away the messy components, the tuning dials, the brass terminals, the ubiquitous wires, into an elegant piece of furniture, or hiding it inside a smoking cabinet, a bookcase, even an armchair with special side panniers. A few months before the Empire Exhibition, the BBC had given the king an absolutely top-end model, presumably on the assumption that royal endorsement could do them no harm. With the doors closed, it blended in with other, antique items of Chippendale furniture in Buckingham Palace, but when open the apparently mid-eighteenth-century neoclassical cabinet revealed its twentieth-century secret.[9] Writing an article about 'Woman's Ways'

PREVIOUS SPREAD This painting by the Scottish artist W.R.S. Stott filled the centrefold of the Christmas 1922 edition of *Illustrated London News*, expansively communicating how wireless might transform the home.

in *The Sketch*,[10] Mabel Howard confidently predicted that, before long, no house would be without a listening-in set. While she acknowledged that these would soon be available to suit all purses, the Chippendale model illustrating her article would have set a household back £170 – roughly the starting annual salary of a male schoolteacher, which might rise to £370 even at the top of his career (and a female teacher could consistently expect 10 per cent less).

Households which boasted such sets were self-selecting in another way. Probably the first British home to be supplied with electricity had been Cragside, built in Northumberland in the 1860s and described by one Victorian periodical as 'truly the palace of a modern magician'.[11] Its electricity came from a hydroelectric generator, which powered not only lighting but also a lift, a rotisserie and a clothes washer. By 1920 all sorts of electrical appliances were available for those who could afford them – toasters, irons, fridges, cookers, food mixers, vacuum cleaners and, of course, radios – but these were only of any interest in homes connected to mains electricity. Their numbers were very small: around 6 per cent in 1919, rising to around 66 per cent by 1939. Even during the later 1920s and 1930s, roll-out was slow, and the retro-fitting of electricity to existing houses often proved problematic and pricey.

Regardless of the type or expense of the receiver, though, an outside antenna or aerial was essential. Paradoxically, these wired contraptions were generally the first sign that a household was now the proud owner of a wireless. In tight-packed Warrington these were strung along alleyways: 'Oh yes, all down the backs, there'd be poles everywhere. They'd use clothes props and brooms and things like that – nail 'em together. As long as it was high up, you'd get a better sound, d'you see.'[12] Aerials frequently appeared in magazine and newspaper cartoons, suggesting how widespread a topic of discussion and source of frustration they were becoming. In one of the earliest *Radio Times* cartoons, a natty 'suburbanite' is standing in his garden in front of a narrow little modern house with a hammock-style aerial mounted between two brick chimneys up on the tiled roof. A visitor has commented on the way that the aerial is sagging, and the owner

It was much more enjoya le to keep silent at Brown's party this year while Aunt Emmelina listened——

Almost every early *Radio Times* carried a cartoon depicting some aspect of the domestic newcomer. This 'then and now' twin image compares an amateur performance of a Victorian parlour song with the new form of entertainment provided by wireless.

replies: 'Yes, I don't know if it is the sun or if it's been warped by the continual "Yes, we have no bananas"',[13] – a reference to the novelty song by Frank Silver and Irving Cohn which had recently arrived from America and already achieved widespread earworm status.

The very first edition of the *Radio Times* had appeared in July 1923 and those who could afford a weekly outlay of 2 pence soon became familiar with its ink-black masthead. This depicted two massive gantries supporting a giant hammock aerial, which hung like empty double washing lines over an awkwardly foreshortened outline of the entire United Kingdom, clearly symbolizing the BBC's ambition to spread the service across the nation. Even its own chief engineer, though, acknowledged that at this stage most wireless listening was characterized not by what you could hear but by whether you could hear anything at all. In the same number, Peter Eckersley wrote an

— than last year when everyone had to keep silent while she sang.

article under the headline 'What are the Wild Waves Saying?', cleverly appealing to listeners who were not only growing in number but also expanding in background, age and levels of technical understanding: the early enthusiasts who knew that wireless telephony was all about electro-magnetic waves, the older generation who would be reassured by his reference to a much-loved Victorian parlour song, and the many, many listeners who complained that, even after endless tuning, what they heard frequently sounded like the pounding of the sea or the muffled roar from a shell lifted to the ear.

The BBC's monopoly had come about at least in part as a reaction against what was perceived as an unregulated 'wild west' model of broadcasting in America – a cacophony of numerous stations and brash advertising. Now, in Britain, the subject of Eckersley's article was in fact a major plan for what he called 'simultaneous broadcasting'. Using telephone trunk lines, this was conceived partly to reduce interference and partly to extend the range of broadcasts so that the whole nation would have access to wireless, but it was ultimately to enable something far more radical – roll-out of the same programmes across the entire country. Today, we take for granted the idea of a national radio network, so it is easy to forget that when the BBC was first launched reaching the whole nation was a technical impossibility. At this early stage the broadcaster was, in effect, a cluster of local stations with headquarters at 2LO, its London station. The earliest of the other stations – Birmingham and Manchester, Newcastle, Cardiff, Glasgow, Aberdeen, Bournemouth and Belfast – were gradually joined by further relay stations in Sheffield, Plymouth, Edinburgh, Liverpool, Leeds–Bradford, Hull, Nottingham, Dundee, Stoke-on-Trent and Swansea. London was to supply the *News*, top artistes and major national events such as the opening of the Empire Exhibition. Simultaneous broadcasting worked in either direction, so listeners across the whole country could hear a programme made in and transmitted from, say, Glasgow (call signal 5SC) or Bournemouth (6BM); but London was always 'first among equals' – increasingly so (and not always happily) as the decade advanced.

The earliest listeners, particularly if they were using crystal radios which had very limited reception, heard local presenters speaking in their own familiar regional accents, announcing events which they might themselves attend in their own neighbourhoods or performances by local entertainers, choirs and bands with whom they were already familiar. Output might have sounded a bit amateur, but then so did much early radio, and it was their own amateur output. On at least one occasion the amateurs literally took over the studios, for example when a group of students invaded the Manchester station 2ZY during rag week, wearing fancy dress and singing – of course – 'Yes! We Have No Bananas'.[14] It must have been thrilling for those listeners who caught this moment of spontaneous pandemonium. On a more regular basis, there was local involvement through competitions, quizzes, charity events and talent contests. Something of this ebullient, two-way spirit was to re-emerge with the comedy talent which helped save the ailing BBC in the years running up to the Second World War (a story covered in a later chapter). For now, though,

Frigidity on the 9.15. A suspected oscillator in doubtful social popularity.

One of the major technical problems which beset many households was oscillation. The BBC attempted to combat this with humour in a free pamphlet.

John Reith's London team assumed greater and greater control, with regional broadcasting replacing its earlier local counterpart and playing often-unwilling second string to the increasingly dominant national network. By late in the next decade *Punch* magazine could run a cartoon in which a comfortably middle-class woman, clearly meant to belong to the southern metropolitan elite, comments to her husband who is tuning the wireless, 'Funny; I could have sworn that was a Regional Cough!'[15]

⁂

While the earlier local patchwork system meant that listeners in different regions might often be hearing different output, tuning in often proved frustrating in similar ways. The problem that caused most tension was a painful form of interference which a single owner could cause their neighbours as far as 30 miles around, simply by incorrectly tuning their own home-made device. This humming, whistling or high-pitched howling was known as 'oscillation' and became the subject of much heated debate. Articles appeared in the press, the BBC published its own pamphlet, and the maverick Peter Eckersley spoke to listeners on air imploring them to avoid such anti-social behaviour: 'Is this fair? Is this British? Don't oscillate. Please don't oscillate. Don't do it', he implored, aware that three-quarters of all listener letters at the time were about oscillation and other problems with reception.[16]

Eckersley was constantly in firefighting mode, attempting to iron out internal glitches which were audible on-air: the ticking of the studio clocks, for example, and other noises made by nervous or ignorant novice broadcasters. Soon a framed notice was pinned up by the microphone in the main 2LO studio, warning guest speakers: 'If you *sneeze* or *rustle papers* you will DEAFEN THOUSANDS!!!'[17] It was not unknown for pioneering broadcasts to descend into utter chaos, as one early BBC employee in Manchester recalled:

> I remember that when we broadcast the Grenadier Guards Band, the players were so crowded together that several had to sit on the piano. We had a unique contrivance for adjusting the height of the microphone to the singer's mouth. The singer stood on a pile of books. One night a tenor in taking a top note also took a step backwards, and there was a terrific crash as he slid under the piano. That ended the solo.[18]

Listeners must have felt a bewildered thrill as they tried to make sense of what was going on. There was nothing routine about listening: the entire experience was so new, so unfamiliar.

That first edition of the *Radio Times* featured a cartoon which, rather patronizingly, explores a common area of ignorance. Entitled 'Our Wireless Village Concert', it depicts a group of people sitting on benches in a crammed village hall which is bare apart from a poster on the wall demanding 'Silence Please'. In the foreground, Mr A. Lotment has tapped the woman in front of him on the back with a request: 'Would you kindly remove your hat, Madame?' The offending headgear is an extravagant black item adorned with a large, curling ostrich feather, but since there is nothing to see except the wireless set positioned, with its large amplifying horn, up on the dais, the joke is on the speaker. It was far from uncommon for cartoons to mock rural listeners, but this notion that wireless was in fact a sort of party trick seems to have persisted: was there actually a person somewhere, speaking, perhaps, from behind a screen or in an adjoining room? Such a notion soon became a plot device for detective stories, such as Walter S. Masterman's *2LO,*[19] published in 1928 (and still available as a reprint for readers in search of real period charm).

A touch more affectionately, a number of cartoons poked fun at older listeners, often dressed in decidedly outmoded Victorian fashion, who continue to be perplexed by the new technology: a plump, smiling elderly lady trying to talk back to the wireless as though it is a telephone;[20] a decrepit old grandpa in headphones saying to his wife, 'Mine's doing [the opera] Faust, what's yours doing?'; tiny, fragile Aunt Lavinia murmuring apprehensively that 'it looks very complicated' as she stands before a solid mahogany wireless cabinet which reaches chin height.[21] Gradually, at least some members even of this older generation were beginning to listen in with enthusiasm. Today, I am reminded of how delighted my own mother was when she mastered email, becoming what she called a 'Silver Surfer' and beginning eagerly (and perfectly competently) to participate in the communication technology which her children and grandchildren already took for granted. In the *Radio Times* Christmas number for 1925 (when my mother was 2) a cartoon juxtaposes two images of dear old Aunt Emmelina, one of her 'listening in', above another in which she is treating the family to her rendition of a traditional parlour song:[22] it's a little unkind but a timely comic suggestion of how domestic entertainment was changing, and, for many, improving.

⁂

The recording of the king's speech was made by a commercial company for its own purposes and not by the BBC itself. All BBC broadcasting in the 1920s was live, although there was extensive use of music played from gramophone records. Output could be fairly random in content and was far from continuous. St George's Day listening was of course centred on events at Wembley, with morning coverage running from 10.30 a.m. to 12.15 p.m. Silence then reigned until 3.30 p.m., when 'a Concert by the Wireless Trio' could be heard, followed by 'The Story of the Stocking Factory' by Helen Greig Souter and 'Across the Rockies by Train' by Agnes M. Miall. At 5.30 p.m. there were 'Children's Stories: Uncle Jeff's Musical Talks with Music by the Orchestra'. Then a further silent interval for forty-five minutes, before the '1st General News Bulletin' and the BBC's drama critic on 'News and Views of the Theatre'. The evening was given over to a rolling event entitled 'St George for Merrie England'. This included 'The Spirit of The Empire and Her Story' reflected in military music, 'Mr Everyman in Stirring Episodes of Empire Building', the baritone Norman Notley singing songs from Shakespeare, a talk on 'The Empire and the League of Nations' by Sir Arthur Salter KCB, and a final rendition of 'God Save the King' at 10.30 p.m. An intriguing item brought this patriotic evening to a close: at 10.35 p.m. the king's speech of the morning could be heard again, only this time it was read out in Esperanto, the artificial language devised in 1887 as a shared medium of communication, which was enjoying its heyday in the 1920s.

Even on such an exciting day, this was a mixed bag. When listeners tuned in, they never knew if they were going to hear an improving talk, a familiar old variety act or something strikingly new. Only a month later the cellist Beatrice Harrison was to perform with a nightingale in her Surrey garden, while the following year the recently established Savoy Orpheans (resident band at the famous London hotel) were to give the British premiere of George Gershwin's 'Rhapsody in Blue', with the composer himself at the piano. Imagine the thrill of the listeners – a young man, perhaps, up in his chilly bedroom, or a recently married couple sitting by the fire in their small parlour – turning on the wireless with no sense that they were about to hear an unearthly nocturnal duet or the syncopated fusion of jazz and classical music which was to become one of the greatest masterpieces of the twentieth century, broadcast live by the BBC.

Already in 1924, two of the five founding executives of the BBC had rushed out books, in which they saw fit to instruct their new public on how to listen to the wireless. In his autobiography, *Broadcast Over Britain*, written at the age of just 35, John Reith gives extensive attention to the matter, usefully painting a vivid picture of the kind of party held by wealthier people and the pitfall of allowing the gadget itself to become the focus of attention:

> We say, 'Come and listen to my wireless set', and we lead our friends into a room where there obtrude on the attention wires and valves and boxes and switches... The attention is distracted by all the paraphernalia and by the tuning preliminaries which ensue. And then we all sit round with our eyes glued to the loudspeaker and come to the conclusion that the sound is metallic and unsatisfying, and that we do not like our music tinned.[23]

In order to avoid such an unappealing start, he recommends a different approach:

> Tell your friends to, 'Come and listen to the Unfinished Symphony', and let the music come on them mysteriously and spontaneously from some invisible source. Camouflage the loudspeaker, hide it behind a screen, in a cabinet, on top of a bookcase – anywhere where people will not sit and stare at it. Why plant the whole apparatus down in the most conspicuous part of the room?[24]

Cecil Arthur Lewis had joined the BBC after distinguished service as an RAF fighter ace in the First World War. His *Broadcasting from Within* is a pacier if rather flimsy read, with an altogether lighter touch. Lewis knew that advertisements often compared listening at home with attending a public venue ('The front row of the Stalls of the Air is Yours...'[25] ran one), but he acknowledged that this very aspect of wireless – the fact it was transmitted into the home – could also divert the listener's attention in ways never experienced in a theatre, hall or concert room:

> In the home a hundred and one distractions call the listener's attention away from the concert. The maid enters with coffee; there are no matches for his pipe, and he goes out to get them; the baby wakes up with a loud and prolonged oscillation (of the vocal cords). He must guard against

> these if he wishes to listen and enjoy to the full. He must help the illusion that this concert is here, in his own home, by switching out the lights and letting himself slip into the mood that is being presented to him.[26]

All this advice was unthinkingly patrician, unapologetically male in tone. It does give a vivid sense of how wealthy listeners who could afford valve sets, which provided proper amplification, were adapting themselves and their homes to wireless. The vast majority, though, needed to use earphones – a very different listening experience indeed.

⁂

Leaf through contemporary publications and you will find page after page of advertisements for earphones, which, almost overnight, had become an essential new accessory. Every market was catered for: lorgnette styles for ladies, lightweight sets for children. An advertisement for a new kind of Electron aerial appears in the 23 April 1924 edition of the *Radio Times*: in vertical format, it shows the members of a single household in a four-storey home, all listening to wireless via earphones, from the son in bed on the top floor down to the fashionable mother or perhaps daughter in an upstairs living room, father in his study, and, in the basement kitchen, the maid, who has neatly fitted her headphones under a perky little lace cap.

Earphones also frequently provided material for comedy. There are staged photographs and touching real-life reminiscences about attempts at sharing: in an intimate gesture of wired connection, courting couples sometimes each held a single earpiece to their ears, which led to daring puns playing on the words oscillation (interference) and osculation (kissing); while one woman later recalled her brother's inventive experiment at resting the headphones in a metal basin, which provided a form of improvised amplification for the whole family. Cartoons often camouflage some uglier truths, though: mostly it was men who owned the headphones and thus men who owned the radio: 'Only one of us could listen in and that was my husband, the rest of us were sat like mummies', recalled one woman from Warrington:

In 1924, when this advertisement for Electron aerial wire appeared, most homes did not have a single radio, let alone one on every floor.

This image from the *Illustrated London News,* November 1923, shows less prosperous people experiencing radio, perhaps for the first time, in an electrical lighting shop which was cleverly cashing in on the new craze.

> We used to row over it when we were courting. I used to say, 'I'm not coming down to your house just to sit around like a stupid fool.' He always had those earphones on, messing with the wire, trying to get different stations. He'd be saying, 'I've got another one', but of course we could never hear it – you could never get those earphones off his head.[27]

For those who could get hold of a set, earphones made listening in a more intensely private and even magical affair than it later became. This sense of the wizardry of wireless was often connected with the concept of *ether* – an invisible medium through which electromagnetic waves were supposedly carried. The word derives from the Ancient Greek for air – not for the air that we breathe or that blows around us as wind (the Greeks had other words for these) but the rarefied element which some still believed filled the upper regions of space. Adam's description of the visiting angel in Milton's *Paradise Lost* as an 'Ethereal Messenger' gives some idea of how it was perceived. In 1899 the eminent physicist and electrical engineer John Ambrose Fleming had suggested that the century to come would be the 'Ether Age', and this concept continued to be cited well into the 1920s. In this era still struggling to come to terms with the cataclysm of the trenches, one surprising believer was Sir Oliver Lodge, a major figure in the development of wireless and a leading public face of science. Lodge's son Raymond had been educated at the same time and the same public school as Peter Eckersley but he had lost his life at Ypres in 1915. In grief, Lodge had visited several mediums and published a bestselling book titled *Raymond, or Life and Death*, about attempting to commune with those beyond the grave. His contradictory endorsement of both scientific and psychic phenomena now seems astonishing, but radio too was still sometimes viewed as at an intersection with magic.

John Reith himself was taken with the timeless properties of ether. The founding of the BBC was not the only great event of 1922. This was also the year in which Howard Carter had discovered the tomb of Tutankhamun, and Reith surely had this in mind when he wrote about the element through which radio waves travelled:

> When we attempt to deal with ether we are immediately involved in the twilight shades of the borderland; darkness presses in on all sides, and the intensity of the darkness is increased by the illuminations which here and there are shed, as the investigators, candle in hand and advancing step by step, peer into the illimitable unknown.[28]

Through earphones, the signal could, at least sometimes, be surprisingly clear (the adjective studio engineers use today is 'bright'). But it was faint, as though the ethereal messenger was speaking from far, far away to each listener's personal inner ear, opening up, as it were, new chambers of the imagination.

In November 1923 the *Illustrated London News* had showcased a photograph entitled 'Listening to *The Children's Hour*' (see p. 54). In this hugely popular and frequently reproduced sepia-toned image, six girls, dressed in loose play dresses (no restrictive Victorian undergarments for them) are clustered together, listening through earphones. They

THE GAME OF RADIO.

DIRECTIONS.

Each player has four counters of his own colour, and starts from one corner of the board, so that if there are four players one will be London, another Plymouth, one Belfast and the other Aberdeen. The object of each player is to call as far away as possible from his own base. Thus a player having started from London, for example, on reaching Belfast may say "London calling." Play then stops, and the scores are taken as follows:

1 for a counter in either of the two towns nearest to your base.
2 for either of the four in the next row.
3 for either of the six beyond.
4 for either of the two nearest your objective.
5 for each counter finishing.

RULES.

After deciding the order of play, each in turn throws the dice and moves a counter in the usual manner along the small red circles. Either a counter in hand or one already started may be moved. An opposing counter in the road you wish to travel may not be passed, so that you must in that case either move another counter or travel in another direction. Observe the advantage of sometimes leaving a counter in another player's way, or going all round the edge for a clear road.

The highest score is of course 20 for finishing with all four counters, but a player need only finish with one if he thinks it is then advisable to call.

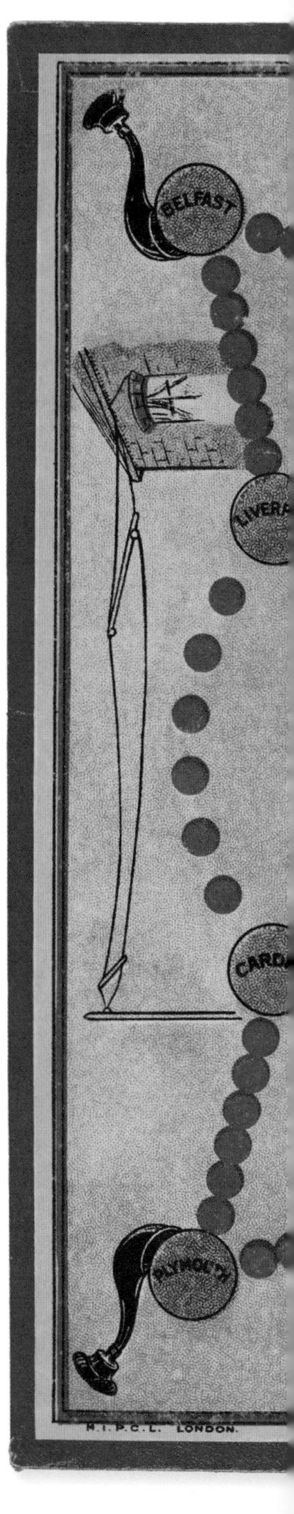

This rare board game must date to late 1924 or a little after, as it depicts 2LO in London, the main regional headquarters and most of the relay stations which had opened across the country by then.

RADIO
THE
WIRELESS
GAME
EDINBURGH
GLASGOW
ABERDEEN
LEEDS
DUNDEF
MANCHESTER
SHEFFIELD
BIRMINGHAM
HULL
BOURNEMOUTH
CHELMSFORD
LONDON
BRITISH MANUFACTURE

have clearly been posed by the photographer, but their intensely interior focus combined with distant gazes give some idea of the impact of wireless. Live broadcasting was, of course, entirely fugitive – no 'listen again', no 'catch-up', no 'downloads' – and this also encouraged listeners to give it their full concentration. The girls in the photograph are Bunty's generation – the very first 'wireless natives'.

Ramsay MacDonald's attitude was not for them (though he eventually changed his mind, and by April 1926 had even agreed to be photographed for the *Radio Times,* sitting in his Hampstead study while listening to a handsome valve set).[29] Children simply took wireless for granted. On the floor, to the right in the photograph, a seventh child looks up towards the others, disconsolate because she is shut out from what they are hearing and knows already that it is the future.

Unsurprisingly, children were often more at ease with the technology than the grown-ups, and this generational divide was soon the subject of cartoons. In real life, wireless was just beginning, perhaps, to alter the family dynamic in their favour (a subject explored in a later chapter). One *Punch* cartoon entitled 'A Child of the Radio Age' clearly caught the eye of the picture editor of the *Radio Times,* where it was later reproduced. It may rather stretch the point:

> Mother: What makes the radio squeal so, Johnny?
> Johnny: Well, mother, if you must know, what you call squeals are really the self oscillations of the thermionic valves, brought about by altering the potentials of the high and low tension batteries and varying the relations of the capacitative and inductive quantities in the receiver.[30]

Meanwhile, less precocious children could now find wireless board games in toy shops. A rare survivor can be found in the John Johnson Collection of the Bodleian Library, in which players attempt to move counters around a course which represents Peter Eckersley's national network, albeit in primary colours. It's not in itself a particularly thrilling game, but clearly the manufacturers were cashing in on public interest. It is certainly a reminder of a time when installing a set in the family home was truly a memorable occasion. *The Radio Yearbook* from 1925 captures something of the excitement:

A 'stocking filler' game like this one, showing a snazzy portable wireless set, must have helped spread the wireless craze to small children.

> When we first get the wireless set from the shop there comes the first thrilling moment when the set is to be operated and the family delighted with music. This, like the wedding day and the first ride on a bicycle, gives a thrill such as we seldom feel in this unromantic age.[31]

In fact, the broadcast of the king's speech at the British Empire Exhibition in April 1924 is credited with inspiring many families to invest in their first set, just as the coronation of Queen Elizabeth II on 2 June 1953 encouraged the purchase of televisions. The kind of communal outdoor listening enjoyed by Bunty and her father was becoming a thing of the past. What listeners now wanted was to hear interesting output on a wireless set with a good, clear signal in their own homes.

FOUR

DAVENTRY CALLING

> My wife releases me [from the shop counter] for the six o'clock News. That is an understood thing. I wouldn't miss the News. I've even neglected the bacon machine for the News.
>
> Mr James, grocer in Bristol[1]

In the middle month of the middle year of the decade, a group of men gathered on the top of a windswept hill in the middle of England. Their heads were tipped back and their eyes directed upwards to a massive new wireless aerial which was suspended between two slender steel masts, 500 feet in height and 800 feet apart. This was the high-power new transmitter on Borough Hill, to the east of the town of Daventry in Northamptonshire. The Postmaster General had travelled here on a special train from Euston, accompanied by the directors of the BBC, for its opening ceremony. As evening approached, they crowded into the small studio where the chief engineer, Peter Eckersley, explained the new technology. Then, at 7.30 p.m. sharp, a red light flashed and the words '5XX. This is Daventry Calling' were broadcast across the nation for the very first time. Lord Gainford, the chairman of the BBC, read out a letter from the prime minister, Stanley Baldwin, declaring that this was truly a turning point, because the new transmitter would bring wireless reception to over 90 per

Winifred Gill in the late 1930s, about the time she and Hilda Jennings conducted their survey into the impact of wireless in Barton Hill, Bristol.

The opening of the Daventry transmitter was a significant milestone in the plan to link the entire country through broadcasting.

cent of the population, a potential audience of 25 million, including those who owned only the most basic sets.

Even better-off listeners with the pricier two- and four-valve receivers had only been able to pick up programmes broadcast from up to 100 miles away, while crystal sets had a reach of just 25 miles. Though quality could still be poor, the earlier patchwork of local stations had favoured listeners in urban areas (which may explain the frequent jokes about the backwardness of country people over all things wireless). In Sheffield, for example, try as they might, listeners often complained of a signal which they compared with the sounds of an insurrection in hell. A cartoon dated a few months before the opening of Daventry shows a well-off couple listening to the wireless. The wife is checking out her *Radio Times* and complains: 'Well, if that's Captain Eckersley speaking, I can't understand a word he says', to which her husband replies, 'My dear, you're looking at the wrong programme. This is a bassoon solo!'[2] The new transmitter, the first of its kind in the world, would in time bring an end to such infuriating reception. Borough Hill was 660 feet above sea level and in the heart of England, so from here it would also be possible to provide the same broadcast service to most of Great Britain – up and down its curving spine, and across its broad middle, from Ipswich to St David's.

From the start, John Reith's vision was for the BBC to be a national broadcaster, which was based on his sincerely held belief that sharing all kinds of output with listeners from every background would have a unifying effect. It had been echoed, with a whiff of wishful thinking, (and a plug for the licence fee), on the front page of the launch edition of the *Radio Times*. In a chirpy editorial, director of programmes Arthur Burrows had drawn a parallel between the function of the new magazine and the universally familiar Victorian railway timetable, suggesting that the BBC would link the country in a similar network:

> Hullo, Everyone! We will now give you the *Radio Times*. The good new times. The Bradshaw of broadcasting. May you never be late for your favourite wave-train. Speed 180,000 miles per second; five-hour non-stops. Family season ticket: First Class, 10s. per year.

The opening of the Daventry transmitter just eighteen months later marked the moment when this mission statement could be put into action, enabling the growing number of would-be listeners to tune in to the same 'five-hour non-stop programmes' (these non-stop hours were actually in the evening, with just an hour of morning music and then a long silence). It was a technical triumph, and the BBC was keen to emphasize its new reach: the *Radio Times* featured photographs of rural folk now able to listen in, along with stories of elderly people trying out wireless for the first time. Commentators frequently pointed out that many disabled and blind people were discovering a boon in radio. But the broadcast to all stations on that night of 27 July 1925 featured a subtler message. This was contained within a specially commissioned poem by Alfred Noyes, who had himself been born in the Midlands and grown up in Wales. His poem evoked foundational myths about the British soil on which the transmitter stood, returning to the popular belief that the town had been called Danetre because Viking settlers had planted an oak on the summit of Borough Hill to mark the very centre of England. At a quarter to eight that evening, listeners heard John Reith's sonorous Aberdonian voice launching into the opening lines. It began like a wireless call signal:

> Daventry calling... Dark and still
> The dead men sleep, at the foot of the hill.

Noyes had breathed life into the eponymous tree and brand-new transmitter as though they were a single, polyphonic, living creature:

> 'Deeper far than the world ye know
> Is the world through which my voices go...'

And then evoked the homes where these voices would be heard:

> I, the sentinel; I, the tree,
> Who binds all worlds in unity,
>
> So that, sitting around your hearth,
> Ye are at one with all on earth.[3]

⁂

Thanks in particular to the commitment of two remarkable and sympathetic women, it is possible to reanimate how the mystical airborne message of the Daventry transmitter was actually converted into grassroots reality. While the Danetree idealistically, and perhaps rather loftily, promised that 'the mind of all the world / is in each little house unfurled', Hilda Jennings and Winifred Gill had access to people living in such homes, and the curiosity and willingness to find out how they were responding to what they heard.

Broadcasting in Everyday Life: A Survey of the Social Effects of the Coming of Broadcasting was commissioned in 1938 as an official BBC publication, so a degree of caution is advisable when reading it, but these forty pages represent a precious oral time capsule. Gill seems to have done much of the legwork, but she and Hilda Jennings clearly collaborated closely on it. It is Gill's manuscript notebooks which are particularly valuable, sometimes because they contain 'off-message' material and sometimes because the passage of time has given certain details a previously unrecognized resonance. Both hugely enlarge our understanding of the impact of wireless in its first seventeen years. The frequently quoted observations by high-ranking churchmen, politicians and dons who mixed in the same circles as John Reith and his senior management team are often revealing and entertaining, but they represent the experience of only a tiny (if powerful) minority. There is precious little first-person testimony from the 80 per cent of the population which made up the working class, giving the research of Jennings and Gill, and Gill's notebooks, all the more value. They are specific to Barton Hill but much would have been true of other parts of the country.

Winifred Gill was a social reformer, artist and craftswoman who had worked before the First World War in the design enterprise known as the Omega Workshops. She was a skilled toymaker and puppeteer, and had already gained a reputation among friends and colleagues for her exceptional ear for mimicry and love of sharing memorable snatches of conversation. This became a real asset when she agreed to undertake the new project. Although her verbatim records can today come across as a little patronizing, it is my belief that she took them down like this in affectionate respect for the cultural and linguistic differences of the people among whom she had chosen to live. By then she was working as subwarden, alongside Hilda Jennings, who was warden, at the Bristol University Settlement, one of a number of missions set up across the country in the late nineteenth and early twentieth centuries, to alleviate lives in deprived inner-city areas. The Bristol Settlement was in the east of the city, in Barton Hill. Since the two were already familiar with local residents, it was agreed that they should focus their small, impressionistic study here, among around 800 households.

Just as the BBC had hoped from the Daventry investment, the three or four years immediately after 1925 saw the people of Barton Hill begin to acquire wirelesses. By the end of the decade a majority owned one, if only on the hire-purchase scheme. Usefully, Jennings and Gill start their study by providing a vivid picture of domestic life in the neighbourhood in the period just before mass uptake.[4]

Once a rural retreat for wealthy Bristol merchants and their families, by the early twentieth century Barton Hill had become a thickly populated district of factories and shops, pubs and schools, with street after street of close-packed terraced houses, some of them in such a poor state of repair that they were soon to be designated for slum clearance. There were a few middle-class or professional residents – doctors, ministers, teachers – but it was largely working-class families living here. Unmarried girls were often employed in the local tobacco factories, in confectionery, shops or offices, while older, married women undertook cleaning work as well as running homes in which there were few labour-saving devices and a wearying routine of cleaning, washing and cooking. Skilled men worked on the railways, in the paper and chemical factories and as engineers; others as dockers, builders' labourers, transport workers, clerks and shopkeepers.

Outside the home, Sunday schools and the settlement or 'mission' provided some social life. Church attendance was so markedly in decline that questions were later asked in the House of Lords about why the local vicar (the splendidly named De Lacey Evans O'Leary) had allowed this to happen in such a populous parish[5] (a fact that Jennings and Gill did not divulge). Married couples might make a weekly trip to 'the pictures', still, of course, silent until *The Jazz Singer* announced the arrival of the talkies in 1927. On summer evenings whole families would bring chairs out into the streets to chat with their neighbours about what was later recalled as a rather limited repertoire of topics – largely births, deaths and marriages

The BBC was eager to establish its relevance to people from all backgrounds and classes. These two rural characters appeared in a 1928 publication entitled *New Ventures in Broadcasting: A Study in Adult Education* and also in *Radio Times.*

among their families and immediate neighbours. Men went to pubs while women spent their time, as they themselves put it, 'talking on the broom handle' at their front doors. When not at school, children played and fought battles from street to street. Without mains electricity, indoor lighting – powered by gas or paraffin – was often feeble. Without central heating, rooms were often cold. There was, anyway, little space in houses, most of which consisted of two rooms upstairs and two down, one a heavily used kitchen at the back, the other a parlour overlooking the street, which was generally kept for special occasions. All in all, it's not hard to see why larger families had tended to spend time outdoors. It is worth noting, though, that some modern historians warn that accounts of pre-wireless leisure activities outside the home should be read with caution, since even the most well-intentioned middle-class observers, such as Winifred Gill in Bristol and the social reformer Seebohm Rowntree in York, often had an instinctive preference for home-based family life.

In the era before wireless, families might spend winter evenings playing cards or listening to a gramophone player, but Gill also hears a good deal about amateur music-making ('when we was courting, of an evening, we'd draw up round the fire and someone would start a song and we'd all join in. Then there'd be hymns, Sundays'[6]). The mandolin was popular, as were the mouth organ and piano accordion, but above all the piano. ('Everyone had to sing in turn – the mother would say "you must hear how our so and so is getting on with the piano".'[7]) This was divided by gender, with one man reflecting that 'boys didn't bother with the piano. Your sister learnt but you didn't – didn't want to...'[8] In reply to Gill's initial question about what people had done in the days before broadcasting, one woman rather glumly replied: 'people did just sit',[9] but others had livelier memories. One of these was Mrs Evans, who fondly recalled her role as in-house entertainer: 'I used to keep the 'ole 'ouse in roars of laughter, sayin' daft things. We was all like that.'[10]

The wireless sets that these households acquired were at the lower end of the price range, but even these were increasingly designed as centrepieces for the home – not to be boxed in as ugly scientific contraptions but proudly exhibited to the neighbours. There were even sometimes complaints about neighbours who played their radios as loud as possible simply to announce the fact that they now owned one. A newspaper story from the other end of the country, Motherwell in Scotland, describes a more generous gesture by a local electrician living in social housing: he had built his own set complete with loud-speakers, and on fine evenings would throw open the window to share the broadcast with his neighbours.[11] For some the object itself became an obsession, as revealed in this complex reminiscence captured by Gill, with its avian coda about the attractions of the domestic sphere:

> One man I know who spent all his time after his work in the public. His wife used to tremble when he came home at night. But now he never goes out on Saturday nights even, he's no trouble. He had a crystal set then a two-valve. He made them himself. He's got two or three sets now – turns on first one and then the other. He used to keep pigeons, but now he breeds canaries.[12]

For most people, though, what they could hear was soon more important than the gadget on which they heard it. Even if, unknowingly, listeners were becoming 'bound in unity' as the Danetree had promised, by the soon familiar invisible ties of the new network.

⁂

The random and varied tone and content of programmes must often have caused bafflement. With such an alarming number of hours now demanding to be filled, the BBC routinely (and sometimes rather desperately) sought out pre-existing musical and dramatic productions and one-off performances to broadcast in full or repurpose for its own output. On the day after the Daventry ceremony, a Tuesday, the main evening entertainment was a concert of music by J.S. Bach, relayed from the Grand Hotel, Eastbourne, with 'Descriptive and Critical Comments by Harold Watts, Mus.Doc (Oxon)'; those who tuned in at the same time on Wednesday must have been a little startled to hear 'Dogs of Devon', a comic opera set on Plymouth Hoe in 1588; on Thursday they could catch a performance of 'Radio Radiance: A Review in Fifteen Beams', performed by a company of West End actors including the soon-to-be superstar radio personality Tommy Handley, along with a rather un-radiophonic-sounding 'Dancing Chorus'.

Each of these might have brought welcome entertainment into homes, but the BBC was also beginning to set in place a growing series of its own regular milestones in what are now called the 'junctions' of the broadcasting day. These have become so familiar that it's easy to forget that in the middle of the 1920s they were a novelty in listeners' lives. Perhaps surprisingly, the news bulletins and current affairs sequences that today form the architectural structure of many networks were not such a feature at the start. It's true that 2LO had launched itself on the world with Arthur Burrows reading a *News* bulletin at six o'clock on the evening of 14 November 1922. It's also worth noting that he did this in stately style, reading it twice, the second time more slowly for listeners who might care to take notes. (In fact, the tempo of all radio talks was extremely slow. Nowadays, we reckon on about 180 words per minute for a scripted talk, while a 1924

BBC memo recommends about half that speed with only 100 words).[13] All this was a clear statement that news was going to play a role in the venture, but the BBC had some outside challenges to overcome. Like all communication revolutions, disruption of the old order had led to some friction. After lobbying from the powerful newspaper industry, which was fearful about losing sales of morning papers, the BBC was, for a long time, permitted only to broadcast news in the early evening. For listeners such as Mr James in his Barton Hill grocery shop the *News* was soon to establish itself as 'appointment listening' – that's to say, a set commitment in his day. Before long, he could also listen out for a daily weather forecast prepared by the Met Office, as well as regular Sunday church services.

Listeners were also beginning to set their watches by the wireless. At the start the sound they had actually heard was the hour being physically struck by one of the announcers on a set of chime bells, but before long a microphone was installed in the tower of Big Ben itself. This unifying and centralizing of time completed the project which had started with telegraphy and the Victorian railways. Peculiar to the BBC, though, was the Greenwich Time Signal or 'pips', which Reith had developed in partnership with the Astronomer Royal, Sir Frank Watson Dyson. These five short tones followed by a longer sixth (the start of this sixth marks the precise hour) now punctuate our internal daily soundtrack, but they have also become emblematic of something larger. At the handover ceremony of the 2008 Beijing Olympic Games the pips were the very first sound heard, an internationally recognized wordless announcement that Britain was to host the 2012 Games. In 1925 listeners were only just becoming familiar with them – some as a reassuring national heartbeat; others, perhaps, as an overbearing reminder of London's dominance. Once again, a *Radio Times* cartoon reflects a kind of history: an elderly man, hearing the signal on his wireless, which sits on a low table in front of a traditional grandfather clock, tuts: 'Dear me, Greenwich five minutes slow! I must write to the papers about it!'[14] There's a similar message contained within a probably apocryphal anecdote related in *The Tatler* magazine, about a small girl who, after reciting her nightly prayer, is overheard gravely promising: 'Tomorrow night at this time, there will be another prayer.'[15]

SOS calls by their very nature interrupted output only irregularly. The very first was an appeal for a missing 6-year-old boy (he was found safe and well) and the concise, unsensational wording soon became familiar: 'Will Mr Black, last heard of in Greenham, please make his way to Blueberry Hospital, where his mother/father is dangerously ill.' These messages were to run for the next seventy years; as time passed, the concept of what did or did not constitute a true emergency sometimes became stretched. A desperate plea for a wet nurse to help feed twins born at a hospital in Norfolk made the grade in 1937; and there must have been a keen ornithologist on the production team during these years, because calls went out both for a customer who had purchased a parrot from a Birmingham pet shop where the seller had subsequently died from what was believed to be psittacosis, and for the safe return of a pelican which had escaped its moorings in St James's Park.[16] As early as 1925 a *Radio Times* cartoon was already poking gentle fun at the new institution: a hapless Bertie Wooster character is pictured on the telephone to the BBC, demanding that an SOS call be put out because he has lost a spat and needs his manservant, who is inconsiderately on his day off, to come home and find it.[17] Even the lightest-hearted crisis (even a cartoon about a crisis) was a reminder of another fundamental change: hearing such broadcasts on the wireless not only collapsed time and space but also began to build a sense of connection between listeners, united in what the political scientist Benedict Anderson was later to call an 'imagined community'. Listeners in Barton Hill and Motherwell were not acquainted with one another and were very unlikely ever to meet, but their growing awareness of the shared experience of listening helped them identify as part of the nation in a wholly new way. A single SOS alert, for example, warning that an outdoor firework had been mis-sold in Folkstone as suitable for indoor use, successfully averted potential tragedy in only one household, but discussion about it extended across the country and merited a short newspaper article as far afield as Hull.[18]

It wasn't much more than a decade earlier that people had been counting their blessings as they sat by their firesides reading about the *Titanic* passengers drowning in the freezing waters of the North Atlantic. The voice of radio amplified this experience now by routinely sharing the dramas of the high seas. Twice a day the Daventry transmitter broadcast the *Shipping Forecast*. The continuation of a wired telegraphic service started in 1861 by Vice Admiral Robert FitzRoy after a tragic shipping disaster off Anglesey, no other BBC icon has produced such resonant poetry. Perhaps finest of all, Carol Ann Duffy and Seamus Heaney were both to choose the elegant, rule-bound sonnet form to reflect on the *Shipping Forecast*. Both end with the liturgical chanting of distant regions, Duffy with a quiet suggestion of the listener's gratitude:

> Darkness outside. Inside, the radio's prayer –
> Rockall. Malin. Dogger. Finisterre.[19]

Both poets crystallize the now familiar feeling of being simultaneously here and elsewhere, in safety and at peril. The residents of Barton Hill and fellow listeners across the country were experiencing this for the first time. When the *Week's Good Cause* became a regular feature of Sunday evening listening in the following year, it was the poor as well as the rich who contributed, their altruism perhaps engaged by a newly expanded awareness of their own relative good fortune.

⁂

Beyond these icons, the earliest programme which achieved this kind of landmark status, and the earliest speech and music programme to become a household name, was *The Children's Hour*. Within weeks of taking to the air in 1922, the director of the Birmingham station (5IT) had dreamed up the idea for a programme specifically aimed at children. Other local stations and London quickly developed their own versions, all comprising a varied mix of live music and talks, drama and stories, news, features, quizzes, outside broadcasts and, in the terminology of the day, 'stunts'. The presenters quickly achieved a sort of celebrity status as 'uncles' and 'aunts' and even an odd 'cousin' or two. These names reinforced the impression that all listeners were united in a single family. In fact, David Davis, who joined the team as an accompanist but later became head of *The Children's Hour*, called it a 'family feeling … the feeling of a small community to which you as a listener belonged and which was coming on for you, personally, every day'.[20]

The programme (today we would call it a strand) was intended for what now seems a very wide age range – children between 7 and 14, the age when most young people started their working lives. There were clearly younger listeners too. *Popular Wireless* magazine featured a photograph of 'Master Eric Knight of Peckham, aged 2½ years', apparently the youngest wireless amateur of Great Britain, who was pictured, along with his teddy bear, wearing lightweight headphones and solemnly staring at the camera as the two of them listened in.[21] Some local stations also developed output for the minority of young people who were staying on at school – series with off-putting titles such as *The Scholar's Hour* – but it was *The Children's Hour* which was most loved, and not only by children. Many

The novelty surrounding listening in was harnessed to sell all sorts of products, from stock cubes to this 1923 advertisement for children's underwear.

cartoons depict older folk who are charmed by its apparent magic, or even monopolizing the headphones so that their grandchildren cannot listen in. An early *Radio Times* cartoon shows a rural grandfather being questioned about the contents of the letter he is about to post. His reply: 'Well, tomorrow be my birthday, so I be just letting the BBC folks know, so as my aunties and uncles can wish me 'appy returns.'[22]

Soon, regular advertisements in newspapers and magazines were indirectly targeting the younger audience, in order to sell their wares. These were both for radio-related accessories, often headphones, and for other goods, such as Oxo stock cubes and the famous Chilprufe brand of underwear (which was worn by my own mother and later by me as a child of the 1960s). Before long, a famous bedtime drink was to provoke an existential battle for the ears of the young audience. It was fought between the new commercial stations and the BBC, and is covered in a later chapter.

As local reception in Bristol was poor, the Barton Hill children and adults heard the 2LO version of *The Children's Hour*, which was now broadcast from Daventry. Billings in the *Radio Times* for the week of the BBC's third birthday in November 1925 give a flavour of what was on offer. On Monday they could hear piano solos performed by Auntie Sophie, along with the chapter of *Tom Brown's Schooldays* titled 'Arthur Makes a Friend', told by Uncle Rex, and 'A Star Talk' by Captain Ainslie. On Tuesday there were trombone solos by F. Taylor, 'The Three Bears', told by Harcourt Williams, and a dose of history titled 'When Charlemagne Was Emperor'. Uncle Jeff could be heard performing piano improvisations on Wednesday while Frank Worthington narrated 'Lost in the Desert', and there was also an item titled 'Temple on the Hill' by C.R. Davy. Thursday promised songs by Ronald Gourley, Uncle Peter telling the story of the 'The Diddle-Diddle Dwarf' and an item on 'Zoo Fashions by L.G.M. of the *Daily Mail*'. 'The Wireless Chorus' sang on Friday, and Rose Fyleman told the story of 'The Price of Four Wishes'. Finally, on Saturday (on Sundays, of course, *The Children's Hour* had its day of rest) there was music by The Octet, Aunt Sophie telling 'The Tale of Squirrel Feathertail', and an unspecified competition.

Once again, this is a pretty random selection and easy to mock, but many recalled the novelty and riches of *The Children's Hour* for the rest of their lives. Now that they had grown used to the wonder of hearing anything at all on the wireless, there was a real thrill in discovering broadcasting intended specifically for them, and all with the particular buzz and jeopardy of live broadcasting. During the earliest years, there was something quite subversive about *The Children's Hour*. In 1924 Uncle Jeff (whose day job was as the BBC's first Director of Music) brought into the 2LO studio his terrier George (whose day job was as a Battersea rescue dog with a marked physical resemblance to Nipper, the famous HMV brand ambassador, and a voice described by his master as mezzo-alto-robusto). There he shared an on-air 'Grand Howl' (a self-explanatory scouting ceremony recently devised by Robert Baden-Powell), in response to which children were asked to send letters describing the reactions of their own pets at home.[23]

The Children's Hour
This fascinating child study in water colours, by a well-known Artist, has been adopted by us for all our boxes for "Brown" Featherweight Headphones. It is also available as a showcard and will be issued as a Calendar for 1926. If you would like one send us a postcard, or apply to one of our representatives.
Brown

Initially unfamiliar, headphones quickly became an essential accessory for most listeners. This image appeared in a Brown sales catalogue and was also offered as a free calendar for 1926.

Early broadcasting for schools is a fascinating story in its own right, but largely beyond the remit of this book because it was heard outside the home. It is safe to say, though, that its output was distinctly different from that of *The Children's Hour*. After a strict day at school, this was an hour (in fact it ran from 5.15 to 6.00 p.m. and was therefore actually 45 minutes, but the title had come from a poem by Longfellow and just seemed to fit) when they could interact with a form of organized chaos which sometimes thrillingly bordered on mayhem. In those early years they could hear the adults sharing noisy backchat and talking across one another in the very way forbidden to them by schoolteachers in most classrooms across the country.

These few years of happily hotchpotch programming were not to last. If in some ways this mini-flagship programme represented in microcosm a version of what the BBC was offering the general audience, it hardly comes as a surprise that by the year of Daventry there were already battles being fought over its content and tone. A year earlier, Reith had already described the programme in his autobiography as 'a happy alternative to the squalor of the streets and backyards',[24] and although the BBC's mission statement to 'inform, educate and entertain' was more than a decade away, this was Reith's personal project from the beginning – and in that order. One of his senior staff members was already reminding the regional Aunts and Uncles that the essential role of children's broadcasting was for 'character building' and urging them to adopt a more dignified on-air presence.[25]

Reith at least recognized that among the widening wireless listenership, there were now many who came from homes of extreme poverty. Cecil Lewis, whose broadcasting alter ego was Uncle Caractacus, seemed oblivious to the fact that most children did not live in homes such as the one in which he had grown up. They might be listening to the wireless, but they were not, to put it another way, the kinds of children whose parents could afford to dress them in Chilprufe undergarments. In a revealing article which he contributed to *The Broadcaster* in 1923 entitled 'The Fun of Uncling', Lewis confessed what a relief it was only to be an uncle of the air: 'I don't hear the cheerful but rather sleepy voices shout "Goodnight" over the bannisters when the evening's fun is over, or the shrill cheers when nurse agrees that there shall be one more game before bedtime.'[26]

Neither this middle-class vision nor Lewis's regular evening sign-off wishing the children a nice hot bath would have been a familiar end to the day for the children of Barton Hill, who lived in houses without nursery maids or bathrooms. We do know from Winifred Gill and Hilda Jennings that many enjoyed it, though we sadly cannot 'listen in' to their observations because the special questionnaires which the older ones filled in have not survived. Perhaps, in the terminology used today to discuss children's literature, *The Children's Hour* provided a window into the lives of others more than a mirror of their own lives.

With the quality of the signal so much improved by the Daventry transmitter, more and more people from an ever-widening range of classes and backgrounds were listening. Whether they felt wholly united within the nation by their listening is a different question. Cecil Lewis's unthinking comment is a reminder of how sharply divided a country Britain was. British wireless listening was about to be politicized for the first time.

This loudspeaker, advertised in September 1925, freed listeners from being tethered by headphone wires whilst also promising to transform family life. The 1920s' term 'radioist' has now disappeared from common parlance.

PRICE 2/-

RADIO FOR THE MILLION

THE RADIO OWNER'S MAGAZINE

Vol. 1., No. 4 September 1927

FIVE

BUILDING JERUSALEM

> It's a funny thing. I've noticed that they don't pronounce words on the wireless like they taught us at school – and ours was considered a very good school for English.
>
> Bristol housewife [1]

On 4 May 1926, after talks between the Trades Union Congress and the government had broken down, Britain's first and, to date, only general strike began. Since the end of the war, life in the coalfield communities of the country had become almost intolerably tough, with the miners' weekly pay dropping from £6 to £3 18*s* over seven years. When the mine owners had declared their intention of reducing pay and increasing working hours once again, the miners had decided to strike, their rallying cry 'Not a penny off the pay, not a minute on the day!' Workers in other heavy industries were the first to down tools in sympathy with the miners. Then the print workers joined them. Once most of the national newspapers stopped appearing, the BBC filled the gap. John Reith himself had broken into a programme of dance music to make a public announcement of the impending strike,[2] and it was on the wireless that news updates could now regularly be heard. The frantic nine days that followed were the first great test of the BBC. For listeners – whether they already owned a wireless

As late as 1927, wireless was still being promoted as an aspirational acquisition for the homes even of 'the million'.

330 RADIO TIMES

BROADCASTING THE NEWS.

LISTENING TO THE STRIKE BULLETINS—

IN TOWN

AND COUNTRY.

Collective listening during the fourteen days of the General Strike provided a further boost to the profile of wireless and to sales.

set, were thinking of buying one or were still resistant – the General Strike was also a turning point.

The immediate change was in the number of bulletins. The original rules which had been imposed on the BBC at the insistence of the news agencies were dropped for the duration, and it was suddenly possible to catch up with the *News* five times a day: at 10 a.m., one o'clock, four o'clock, seven o'clock, and 9.30 in the evening.[3] The routine which was to become so familiar – the punctuating of personal schedules by regular news updates – was in itself a novelty.

In a population of 45 million, only 2 million people held licences, and for them and their households this was surely a great moment of early-adopter self-congratulation. Almost everyone, though, had access to wireless news. Many listened to the public loudspeakers which were popping up in all sorts of localities. The owners of radio shops, quickly recognizing that the national crisis might also provide them with a marketing opportunity, ran wires out of their windows and connected them to amplifiers on the streets; town halls set up notice boards so that the public could read the latest BBC bulletins as they were swiftly transcribed by teams of shorthand typists.[4] Across the country there were resourceful local solutions to the news blackouts: two sixth-formers at a grammar school in a small coal-mining town in North Wales, where radio ownership was still rare, were permitted to sit listening to the wireless in the headmaster's study, taking turns to transcribe the BBC bulletins and then typing them up and posting them in the windows of the local stationers.[5] As the social reformer Beatrice Webb succinctly noted in her diary: 'The sensation of a General Strike, which stops the press as witnessed from a cottage in the country, centres round the headphones of the wireless set.'[6] For these nine days, there was a palpable wartime spirit but with a fundamental difference: there was no Kaiser Bill, no common enemy. On the contrary: this was an internal battle, a stark reminder of the division between rich and poor, which some even believed might become a revolution. Strikers listened in town squares with a growing sense that the BBC was on the side of the establishment. Meanwhile teams of middle- and upper-class volunteers who were responding to the government's call to keep essential services running clustered around loudspeakers set up on flatbed lorries. For these 'plus-four volunteers'[7] this was a patriotic gesture. Many young men down from the old universities saw it as a 'rag', a 'lark for the sake of their country',[8] and even hoped it might enable them to fulfil boyhood fantasies of becoming an engine driver for a day. For women, there was a reminder of the opportunities which the war had provided, volunteering in canteens, as messengers or telephone operators. They knew this was only a temporary change to their normal way of life, while the strikers were fighting an existential battle, its long-term consequences unknown and quite possibly entailing unemployment and more desperate poverty. The difference in mood and atmosphere

This Marconiphone Universal Baby Crystal Receiver from around 1924 was the kind of wireless on which many listened to news during the General Strike.

was extreme: far from being a single, collective experience, public listening often reflected these opposing political allegiances.

In private, some who had previously been resistant to buying a wireless now succumbed. The Bishop of Winchester later confessed that, after 'manfully resisting the numerous appeals of wireless', he now made the decision to install a set.[9] The moral judgements implicit in the words 'manfully resisting' and 'numerous appeals' give a sense of the disdain with which he had previously viewed wireless (or believed a man of the cloth should view it), but keeping abreast of the news justified the purchase. Even the poorest listeners with their cheap crystal sets were more likely to catch bulletins, because Peter Eckersley had arranged for the signal to be specially boosted so that those in remoter areas would still be able to listen in their own homes or in those of slightly more fortunate neighbours.

As in every crisis since 1926, whether the Second World War or the Covid pandemic, there was also a sense of reassurance to be found simply in listening to the regular programming which continued to be scheduled around the *News*. This experience sent down the first roots of what has now grown into a warmly cherished belief that if the BBC is still broadcasting then the end of the world has not yet arrived. Projecting forward, this is now represented by the possibly urban myth known as 'the Droitwich signal', named after the BBC's most powerful transmitter, which was erected in 1934. It is said that if the commanders of British nuclear submarines believe that the country has suffered a nuclear attack they are instructed, along with other contingencies, to tune into BBC Radio Four Long Wave in order to discover whether the United Kingdom is still functioning.[10] If they cannot receive the signal, they must open sealed orders that may authorize a retaliatory attack. Imagine the commander's relief: the threat of nuclear apocalypse averted by the reassuring voice of Clare Balding's ramblings, Evan Davis anatomizing the national economy, or Neil Nunes trailing events in Ambridge just before the seven o'clock News. Back in 1926, a sardonic but perceptive journalist from the *New Statesman* suggested that what then seemed a novelty in fact had a historic pedigree stretching back many centuries before the arrival of wireless:

> A good many people, I am sure, were charmed when, on the first day of the first General Strike that this island has known, they turned to the wireless for news of what was happening to themselves and their civilisation and found themselves, having heard all the news that was fit to transmit, invited to listen to a talk on 'holidays with ants and grass-hoppers'. Since Drake played out his game of bowls while the Armada was approaching, there have been few things more pleasurably characteristic of the English attitude to a crisis.[11]

News was the critical attraction, though, and there were some creative approaches to catching it. Decades later, an elderly man recalled 'the little home-made crystal set, which worked lovely with the iron bedstead for aerial and the gas stove for earth, and which told me and my wife (each with one earphone to the ear) what was really happening'.[12]

As a statement of faith, the closing words are poignant, because we now know that the BBC was not, in fact, impartially sharing news of 'what was really happening'. Many historians have anatomized the BBC's failure to portray the realities of working-class life and the spirit and solidarity of the strikers, or to give anything like equal airtime to representatives from either side. Some have also shone a spotlight on the tightrope which the broadcaster had to walk in order to maintain a degree of independence in the face of pressure from Stanley Baldwin's government, and in particular from his ragingly anti-Trades Union Chancellor of the Exchequer, Winston Churchill. Above all, David Hendy's recent account, brilliantly pieced together from original documents, provides a lucid and thrilling day-by-day, sometimes hour-by-hour, narrative. It is also scrupulously fair-minded in acknowledging that the BBC's whole future was in jeopardy. This was the prototype of many psychodramas in the decades to follow, but the counterfactual stakes in 1926 were particularly high: if the government had been more deeply dissatisfied, it might have refused to grant the royal charter which was currently under discussion, and the BBC as we know it would have died in its infancy.

When the strikers finally capitulated in the late morning of 12 May, it was once again John Reith himself who made the on-air announcement. By teatime, listeners could hear him reading out statements from the government and from the TUC, as well as his own coded acknowledgement of the shortcomings of the BBC's performance, and assurances that one day the full behind-the-scenes story would be told. What followed was a quietly sensational three-minute performance[13] which foreshadowed something of the role that the BBC, at its best, would later assume.

To the accompaniment of a small live orchestra, Reith recited all four stanzas of the poem by William Blake now known as 'Jerusalem', culminating in its rousing promise of a better future for all:

> I will not cease from Mental Fight,
> Nor shall my Sword sleep in my hand:
> Till we have built Jerusalem,
> In Englands green & pleasant Land.

Written in 1808, the poem had been little known in the century that followed. It had only recently become popular, after the Poet Laureate, Robert Bridges, had included it in a 1916 anthology designed to raise morale during the First World War, and had then asked the composer Sir Hubert Parry to set it to music. With an emotional intelligence we do not always associate with him, Reith was tapping into the poem's capacity for widely diverging interpretations. On the one hand, Blake had been an outspoken supporter of the French Revolution and a fierce critic of what he considered the enslavement of the masses in the cotton mills and collieries of his day. Miners and TUC members gathered round public amplifiers or their own cheap crystal sets might surely count him among their number and feel a degree of comfort? On the other hand, might the hearts of many a High Tory, listening on a 100-guinea valve set, not be warmed by the poet's optimistic paean to a universal humanity rooted in English soil? This was a poem which was already embraced across the political divide: religious in tone though not strictly speaking a hymn, sung by both Labour and Conservative supporters, it had been adopted by the women's suffrage movement and, just two years previously, by the Women's Institute. Listener allegiance might be sharply divided but they were united by the same broadcast.

The BBC's General Strike coverage had been very good indeed for its public profile, although in the weeks that followed some wrote to complain of its biased coverage, and the *Radio Times* also sought out the views of a number of public figures. The response of the passionately pro-strike Labour MP Ellen Wilkinson is worth reading at length:

> The attitude of the BBC during the crisis caused pain and indignation in many subscribers. I travelled by car over two thousand miles during the strike and addressed very many meetings. Everywhere the complaints were bitter that a national service subscribed to by every class should have given only one side during the dispute. Personally, I feel like asking the Postmaster General for my licence fee back as I can hear enough fairy tales in the House of Commons without paying ten shillings a year to hear more.[14]

In fact, increasing numbers of listeners were now finding the necessary 10 shillings to pay for a licence – uptake in 1926 increased by 32 per cent – but wireless was in some ways highlighting the very divisions between rich and poor which made supporters of the Strike, such as Ellen Wilkinson, so indignant.

Back in July 1924 the *Birmingham Evening Despatch* had carried a story under the quadruple headline, 'Crystal Set Irony – Savoy Band in Miners' Hovel – Family Misery – Father, Mother and Three Children in One Room'.[15] It described the atrocious housing standards in the coalfield villages surrounding the city of Durham: here were families living in a single room in conditions which it declared were reminiscent of the thirteenth century. The only evidence of the twentieth century was an aerial in the back yard and a small crystal wireless set in the window.

In an affecting article, which culminated in the mother declaring that all she longed for was a second room, the words 'Savoy Band in Miners' Hovel' do the heaviest lifting. The Savoy was one of London's most luxurious landmarks, famously boasting lavish bedrooms, hot and cold running water in the ensuite bathrooms, electric lighting throughout, 24-hour room service and two resident dance bands. The vivid contrast with the miners' homes prefigures the industrial unrest which was soon to come. It also exposes what must often have been a startling dissonance between the programmes which wireless brought into homes and the lives of the people who heard them, making listening at one and the same time an enriching and disorienting experience.

John Reith's passionate dream that radio would bring about a kind of national levelling up was increasingly represented in cartoons. A socially complex example was published a few years later in *Punch*: two working men sit in a cluttered kitchen; the woman of the house is doing the washing up and a dog sleeps by the fire. The curtains are drawn, the atmosphere cosy as they listen to a concert broadcast on the wireless. The cartoon is entitled 'The New Critics', with one man saying to the other: 'The *pizzicato* for the double basses in the coda seems to me to want body, Alf.' Like many *Punch* cartoons of the time, it reveals a patronising attitude towards working- and lower-middle-class aspiration – in this case, the appreciation of serious classical music. Even so, this message is softened by a more comfortable suggestion that radio might help bring about modest upward social mobility.

Such optimism was wholly absent from other cartoons, even those appearing in the *Radio Times*. Only a few weeks after the General

THE NEW CRITICS.
"The *pizzicato* for the double basses in the coda seems to me to want body, Alf."

This 1932 *Punch* cartoon both warmly celebrates radio in the home and discloses contemporary class assumptions.

Strike had ended and in prime position just inside the front cover, the 16 July 1926 edition showcased a single brilliantly expressive image about class division. The scene is a large office, furnished in masculine opulence with buttoned leather armchairs and a massive partners desk. In the foreground stands a middle-aged woman: hair scrunched back, she is painfully thin, clothed in a drab dress and shoes worn down at the heels. She holds a mop in one hand and a bucket in the other and has just pushed back the heavy carpet to clean the marble floor

"We are now taking you over to the Savoy Hotel."

This cartoon, with its barbed observation of the stark divisions in British society, appeared in the *Radio Times* in July 1926, surprisingly soon after the end of the General Strike.

beneath. It is the look on her face which tells the story: a startled yet weary resignation at the absurdity of the words she is hearing broadcast from the conspicuously pricey radio sitting on a cabinet: 'We are now taking you over to the Savoy Hotel.'[16]

The music of the two bands of the Savoy Hotel and many other similar orchestras based in grand London venues brought immense pleasure into millions of lives, perhaps particularly the lives of the very poor. Why otherwise would those who lived in households where there were scant if any savings at the end of each week have sacrificed other essentials or small luxuries to purchase a basic crystal set and a licence? With no running costs, the wireless brought free entertainment into homes. Across a single week (though not, of course, on Sunday) in June 1926, exactly a month after the General Strike's end, listeners could hear Alex Fryer's Orchestra, Jack Payne's Hotel Cecil Dance Band, the London Radio Dance Band, Jay Whidden and his Midnight Follies Dance Band from the Hotel Metropole, Al Davidson's Band from the Café de Paris, New Verrey's Dance Band from the New

Verrey's Hotel, Frank Westfield's Orchestra, and Kettner's Dance Band from Kettner's Restaurant.

Since the opening of the Daventry transmitter, wireless programmes were increasingly coming from the heart of London rather than from the regional outposts. Listeners often heard on-air the address of the BBC headquarters, which was in Savoy Hill, just off the Strand and right next door to the hotel. The tone of the BBC seemed naturally to belong in such hotels or in the homes of the cultured, highly educated metropolitan elite.

It is often assumed that Reith and his senior management team had all attended public school and Oxbridge. Remarkably, not one of them was university-educated. Reith and Eckersley both had backgrounds in engineering, Cecil Lewis had joined the RAF straight from school when war was declared, Stanton Jefferies had studied at the Royal College of Music, and Arthur Burrows was in fact the son of an Oxford college porter and had made his way up through local journalism and as an employee of Marconi. Even so, Reith's cultural assumptions and uncompromising personal mission carried absolute sway. These soon covered every aspect of what we might now call the public image of the BBC, a 'house style' which was far removed from the lives and lifestyles of the vast majority of listeners in their own homes.

After the improvisation and informality of the early years, it had been decided by 1926 that BBC announcers across the country should not only now be anonymous but also more formal in what they wore and their on-air style. The rule about evening dress, which seems particularly quaint for an aural medium, was based on the idea that, at least in the evening, announcers might regularly be greeting artistes and other guests who would be similarly turned out, so this, it was argued, was surely no more than good manners. Their speech was actively intended to set the highest possible standard. Even regional announcers, Reith argued, should be 'men of culture, experience and knowledge' who would 'build up in the public mind the sense of the BBC's collective personality'.[17] This personality was, of course, based on upper-middle-class values, diction, accent and pronunciation. Under the auspices of the newly formed BBC Advisory Committee on Spoken English, work began in 1926 on two influential pamphlets. The first was to establish standard pronunciation of what were called 'Doubtful Words', and the second was on English (*sic*) place names. Both emphasized that this was largely for the sake of on-air consistency and would provide a sort of aural common denominator which was not in any way intended to cancel local custom (which the BBC, at least in principle, favoured). A telling comparison was made with (male) dress codes: 'The kilt is as conspicuous in Piccadilly as the silk hat upon the moors: there are, however, occasions when a black tie is considered suitable by all classes.'[18]

The author of the introductions to these pamphlets was the BBC's linguistic adviser Arthur Lloyd James, a Welsh-born professor of phonetics at the University of London, who was later to gain notoriety for murdering his wife with a fork and poker in a fit of insanity. In the mid-1920s, he was sympathetic – sometimes even humorously so – towards regional differences of pronunciation, but he was unapologetic about the BBC's ultimate criteria for decision-making. When it came to pronunciation of the place name Newcastle, for example, he knew just where he stood in the choice between the short *a* and stress on the second syllable, as pronounced by those who live there, and the long *a* and stress on the first syllable used in the south:

> The only solution of this problem is to regard the Northern variety as one peculiar to Northern England, and since Announcers are not required to use Northern English, to recommend the Southern variety for broadcasting, at any rate from Southern stations.[19]

Now that the name Daventry was so frequently heard on air, its contested pronunciation came up for particularly lively and unintentionally revealing debate:

> There are people who believe that they are doing a service to their country and its language when they advocate that a word looking like Daventry must sound like Daintry. It is difficult to see what would be gained, but it is equally difficult to persuade these people away from their point of view. If it is considered vulgar to say 'haint we' for 'haven't

we', why is it not vulgar to say 'Daintry' for 'Daventry', for the principle governing the loss of the 'v' is the same in both cases? But 'haint we' cannot appear in print, unless it appears as emerging from the mouth of an uncouth speaker; its use by the educated is a joke to-day, though there doubtless was a time when it was quite respectable...[20]

Lloyd James continued in this vein for several sentences before pointing out that the decision had, in effect, already been made by the BBC's own regular on-air use of the phonetic pronunciation of 'Daventry'. In this he was reflecting the fact that the broadcaster was swiftly becoming the final arbiter of standardized pronunciation.

The breezy self-assurance of these pamphlets, their paternalism, is startling today. In the mid-1920s, though, when the majority of the population left school aged 14, the daily flow of a uniform version of 'correct' speech on the wireless seems frequently to have been perceived as beneficial. One man from Barton Hill certainly suggested this: 'Pronunciation of difficult words on the wireless is a great help, [it] gives confidence ... it largely does away with inferiority complex[es] in people who don't go out much and who would be right out of things.'[21] A woman observed: 'I notice how they say it on the wireless and then I judge people according... Children do talk older than they did.'[22] Mrs Bissenden commented: 'The wireless help you to say words – you hear them, and then afterwards, you use them yourself. And it makes the children better able to say what they want to.'[23] And from those involved in education there was a sense of satisfaction in the impact on children: 'Boys know that there IS another way of speaking. Where the house is backward, the boy is probably backward too – [he] doesn't listen properly to the wireless. But the intelligent home becomes more intelligent.'[24]

In areas where local educational authorities had funded wirelesses in schools, there was soon a regular strand of programmes called *King's English*. Feedback from schoolteachers was positive – 'no more foight the good foight'[25] observed one schoolmaster about the benefits on hymn-singing – but there was a tacit consensus that it was only the individual words spoken by pupils, rather than their full sentences, that altered, and that any changes (perceived and described as 'improvements') in the classroom very rarely spilled out into the playground. Arthur Lloyd James went so far as to call this divide between official and unofficial speech a kind of 'bilingualism'[26] and to defend it as a useful informal qualification for those with aspirations in the job market. Jennings and Gill observed a similar phenomenon:

> in ordinary conversation in the home, especially among the older people, the local colloquial mode of speech with its native raciness holds its own. In some instances, whether as a result mainly of broadcasting or of education in the schools, there appears to be a conscious use of two distinct modes of speech in the homes and at business or for social purposes.[27]

The overall benefits of feeling more articulate, more able to express oneself when speaking aloud, more confident when reading in public, were much enjoyed by the generally self-assured grocer, Mr James:

> When it comes to diction: I have a bit of an ambition to do a bit of public speaking myself... My chief problem is the reading of the Holy Scripture but do you know I noticed that these announcers sometimes make mistakes in reading their manuscripts and I say to myself, there, you see, if they can make mistakes don't be so silly and upset when you stumble in reading.[28]

Today, the BBC's specialist Pronunciation Unit does exceptional work in its role as advisor to announcers and newsreaders on any word, name or phrase in any language, especially those which arise day by day in topical stories, but it does not impose a uniform pronunciation on all presenters and contributors. A diversity of accents, styles and backgrounds is, in fact, agreed to be desirable. In the mid-1920s, only the voices of a very small minority of the population were routinely represented in any of the main on-air sequences.

⁂

Listeners had been able to hear working-class voices from the very beginning but these were almost invariably performing comic turns. Helena Millais appeared on 2LO when it was still Marconi's London station, three weeks before the BBC began to broadcast. By good fortune (and unusually for these early years) she can still be heard today because she was also to become Britain's first recorded broadcast comedian, laying down a handful of her acts on gramophone records in 1928.[29] Trained as a classical actress, she created a cockney wireless persona called Our Lizzie, whose apparently meandering (but in fact elegantly structured) monologues she herself wrote.

With vowel sounds which ran roughshod over the recommendations of the BBC's Advisory Committee on Spoken English, Lizzie's patter began: ''Ello me ducks, 'ere I am again wiv me old string bag', and ended with the advice to be good 'and if you can't, mind no one sees yer'. Each 'comedy fragment from life' focused on her own imaginary family, often alluding to clashes in class aspiration, such as when 'our Bert' explained that he drank his tea out of a saucer because he couldn't find a way of drinking from the cup without the spoon going up his nose. Her act has aged well in spite of the many ways in which, like much British comedy, it reinforced the idea that people with accents or modes of speech which differed from the approved version were worthy only of affectionate ridicule. Another of the recorded episodes, entitled 'Our Lizzie Listens-in', provides a timely reminiscence of early wireless listening itself:

> My neighbour's husband Mr Green, he was the first to listen-in down our street, and I'll never forget the night he brought it home. I was in their back kitchen when in he come and he put a box down on the kitchen table... 'There,' he says, 'What do you think I've got there?' I says, 'White mice'. He says, 'Don't talk silly, that's a crystal set!' I says, 'Go on, I never knew you told fortunes!' 'No,' he says, 'That's what you listen-in.' 'Well,' I say, 'It's you what's talking silly, no one could get into there to listen to anything...' He pushed me into an armchair and put a pair of goggles on me ears. 'Now,' he says, 'Listen!... what can you hear?'. I says, 'Nothing', and just then I heard someone say '2LO calling', so I says, 'Two 'ellos to you too!'. Mr Green says, 'Who are you talking to?' I says: 'I don't know, but he sounds matey...'[30]

Not until the following decade could working-class listeners hear real voices such as their own on air. A key pioneer in this was Olive Shapley, a young Oxford graduate who had joined the influential Manchester station in 1934 as an organizer on *The Children's Hour* and was soon promoted to assistant producer. At a time of great formality in broadcasting, it wasn't easy to achieve naturalness. On one early live transmission, Shapley had to write big signs in the studio reminding a group of Durham miners 'not to say "bugger" or "bloody" on air',[31] but she was unwavering in her commitment to showcasing the unmediated voices and experiences particularly of industrial workers. Soon she developed an interest in what we would now call documentary features and took advantage of the newly developed mobile recording vehicles to interview people in their own homes. Her first programme was on shopping; she went on to make documentaries on canal workers, long-distance lorry drivers, homeless people and miners' wives. The finished programmes were the forerunners of documentaries heard on radio today, and some seen on television: they had a scripted introduction and links, but the unrehearsed, spontaneous contributions from her interviewees were unlike anything that had been heard on British radio before. These, though, were still a decade away.

Back in 1926, the BBC's rather haughty authority seemed to be transforming many other aspects of society to a remarkable degree for an institution which was not yet five years old. One 1926 *Radio Times* cartoon, for example, suggests that the metropolitan broadcaster was already supplanting the ancient lore of the countryside and the traditionally respected knowledge of its residents. As evening approaches, a laconic interchange between two rural characters runs as follows:

> 'Sky be very bad to-night, Jarge. What do that mean?'
> 'Can't tell 'e, 'Erbert. Us'll 'ear it on wireless later.'[32]

In real life it was not only national politicians such as Ellen Wilkinson who challenged such authority. Articulate Mr Massey in Barton Hill was unusual but not alone in voicing his frustration: 'You see, you can't put questions to the wireless.' He was also ahead of his time in expressing his opinion that the balance of voices on-air was rather skewed, and in suggesting how he would like this changed: 'a little more from the ordinary man's point of view instead of professors. One speaks as he sees, another as he knows it from a special standpoint, which is often not applicable to ordinary life.'[33] The later

Rich and poor, urban and rural, young and old: this elegant pair of drawings from a Marconi pamphlet on *The Art and Technique of Broadcasting* reinforces the message that wireless listening is for everyone.

programmes of Olive Shapley must surely have pleased him, and many of the men and women who had been involved in the General Strike and been disappointed by the version of events they had heard on the wireless.

Mr Massey spoke of 'the ordinary man': in official BBC pamphlets and elsewhere, the listener was routinely referred to as a 'he'. What, though, of the 'ordinary woman'? A few might still be rejecting wireless – Our Lizzie's neighbour Mrs Green, for example ('she wouldn't listen-in 'erself, said it tore 'er 'airnet, well fancy wearing an 'airnet, you know, she's got thoughts above 'er station'), but increasing numbers were beginning to share Lizzie's own experience and attitude:

> Our Bert made us a set, out of a cigar box and a few tintacks. Mind, he broke a few things while he was doing it – took me broom 'andles and me clothes line and broke a chair and lost the 'ammer and fell out of the kitchen window, but he done his work well. So, if you ever want to listen-in, ducks, come round to Lizzie's 'ouse and have a real good time.[34]

It was women, after all, who spent most time at home and whose daily lives might be most affected by wireless.

Radio Times, November 16, 1934.
THE
RADIO TIMES
503
MONDAY
767 kc/s 391.1 m.
Midland Regional
12.0 An Organ Recital by
1.0 A Program
3.0
9.40 'The News'
Regional
LONDON REGIONAL
877 kc/s 342.1 m.
12.0 THE MIDLAND STUDIO ORCHES
Directed by Frank
March, Sounds of Peace...
Overture, The Italian in
Concert Waltz
Gypsy, sing for me
Serenade, O sole mio........di Cap
A Funny Fellow
Scottish Re
10.15 THE DAILY SERVICE
Time Signal, Greenwich, at 10.30
10.45 The Mother's Health—2
By a Doctor
Selection,
Minuet...
Musical S
Waltz Mec
Northumb
National Pro
5.45 Chamber Music and Songs of John Ireland.
7.30 Olga Haley (mezzo-soprano)
9.30 Concert relayed from Felixstowe
Monday
12.0 Tom Jenkins (organ)
1.15 Northern Studio Orchestra
3.45 Buxton Municip Orchestra
...BAKE FOR 10 MINUTES
THIS IS THE FIRST NEWS BULL
THIS IS THE NATIONAL
hullo everybody -- this
WILL FOLLOW IN ABOUT TWO MIN
The cuckoo builds it's ne
2D
WOMAN'S
BROADCASTING NUMBER

SIX

THE 'FLAPPER ELECTION'

> You will be pleased to hear how much I have enjoyed your news of the air race. I have enjoyed equally the three-minute intervals, which have given me time to reach the kitchen and baste the joint for dinner.
>
> Housewife in a letter to the Marconi Company, 1922[1]

The forecast for Thursday 30 May 1929 was bright and clear with a light breeze. Fine weather traditionally brings out the voters, but a newly enfranchised cohort of women needed little encouragement to set off for their local polling stations. British women had been fighting for the vote since the 1830s. In 1918 the battle had been won, at least for women over the age of 30 who met certain property qualifications. Ten years later, in 1928, the Equal Franchise Act had finally given women the same voting rights as men. This was a day of monumental importance, though the newspapers had come up with a name which cut it firmly down to size: the 'Flapper Election'. Back in 1920, *The Times* newspaper had reported on the lecture of an eminent Scottish doctor who had casually referred to 'The social butterfly type ... the frivolous, scantily-clad, jazzing flapper, irresponsible and undisciplined, to whom a dance, a new hat, or a man with a car, [are] of more importance than the fate of nations.'[2] The fate of the nation might well be in peril, the doctor implied, if women were to

This arresting *Radio Times* cover from November 1934 was one of a number of editions devoted to women listeners. In its bold graphic design and colour, it emphasized the modernity of the medium.

This 1929 image from *The Graphic* (which clearly had far laxer editorial standards than *Radio Times*) suggests that 'bright young thing' Hermione Bonsonby-Peaton is far too hung-over to appreciate a broadcast on the benefits of drinking milk.

be given sway in the public domain. The word 'flapper' may derive from a northern English name for a teenage girl who has not yet put her hair up and whose plaited pigtails flap down her back; or from a far older slang term for a young prostitute. Neither suggests a complimentary attitude towards the 15 million grown women who were now eligible to vote.

In private life, too, the image, role and daily experience of women were contested. It's almost as tricky to generalize about the women of Britain in 1929 as it is about the men, but most married women did share a common status: whether urban, suburban or rural, from whichever region, class or background, they were housewives, and therefore the main daytime consumers of wireless in the home. There were, of course, some men tuning in too – shift workers and an increasing number of the unemployed, particularly as the impact of the Wall Street Crash began to play out over the following years – but since women made up the vast majority of daytime listeners, programmes were largely designed for and aimed at them.

⁂

A month before the founding of the BBC in 1922, a magazine entitled *The Broadcaster* had casually stated 'that women will not be interested in the mechanical side of broadcasting is a fact, for the natural indifference of the fair sex to any knowledge of "what makes the wheels go round" is inevitable.'[3] Women were also often depicted as ignorant saboteurs of male attempts at constructing wirelesses: a housewife in Bovey Tracey in Devon, for example, was reported to have been found hanging her washing from an aerial.[4] By 1929, though, the idea that radio was exclusively a hobby for men was gradually giving way to marginally broader-minded attitudes, though even these were laced with casual sexism. The (male) writer of a radio column in *The Sketch*, a popular illustrated magazine, was keen to point out that his postbag was increasingly filled with letters from women, many of whom were not only buying ready-made radios and installing and operating them themselves, but also building their own sets and asking questions 'that revealed no small knowledge of the general principles of the subject'. He began to blot his copybook in the penultimate paragraph:

> I can see no reason why any woman should not work a wireless set quite as intelligently as a mere male. The majority of modern wireless sets, in any case, require very little controlling, for it is generally a case of just pressing one switch to put it on or off, or revolving one dial to tune in other stations, but even the ones with rather more elaborate controls will easily respond to the very nimble fingers of the fair sex. Tuning is an art that requires not a heavy hand, but a most delicate touch, and there is really no set to-day which is beyond the control of the modern Miss.[5]

His closing words, though, were a wireless version of common contemporary tropes about powerful women and their henpecked husbands: 'It has, of course, been very rudely said that no woman likes a loud-speaker, as she cannot answer back – but then she can always switch off and so have the last word!'

TEN YEARS OF LISTENING

IN 1922 I GOT INTERESTED IN WIRELESS

IN 1923 MY CRYSTAL SET BEGAN TO FUNCTION

& IN 1924 IT PASSED THE EXPERIMENTAL STAGE INTO THE DRAWING ROOM.

WITH 1925 CAME THE GLIMMER OF VALVES,

& IN 1926 I GOT MY FIRST LOUD-SPEAKER.

IN 1927 I GOT A LOUDER-SPEAKER

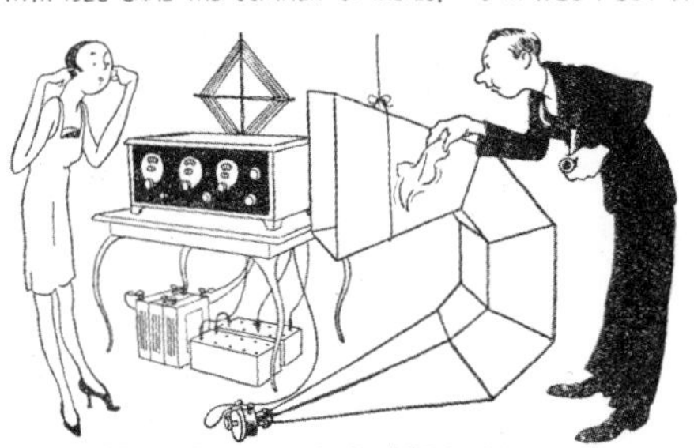

& IN 1928 A REALLY-LOUD-SPEAKER.

IN 1929 I ADDED A (NOT VERY) PORTABLE.

IN 1930 I SWITCHED OVER TO AN ALL-MAINS SET

& IN 1931 A RADIO-GRAM.

NOW IN 1932 I HAVE GOT EVERYTHING – SO I'VE TAKEN OUT A LICENCE!

Married men and women at least tended to agree when it came to deciding where in their homes to put the new wireless set. The parlour was the highest-status room, in the front of the house, and traditionally reserved 'for best' and for Sundays, for rituals of birth, marriage and death, or for greeting the outside world. Even the new social housing which had been constructed in the 'Homes Fit for Heroes' post-war building boom often had parlours.[6] The 'ordinary' women consulted by a 1919 Women's Housing Sub-Committee of the Ministry of Reconstruction had quite expressly stated their desire for a parlour, though the space that this took up often meant that the kitchen at the back of the house – where they generally spent far more time – would be smaller and pokier. The parlour was the room in the house which had commonly boasted an upright piano in the Victorian period and gained a gramophone player during the first two decades of the twentieth century. Now the advertising departments of radio manufacturers and the editors of wireless magazines, including the *Radio Times*, went into overdrive, promoting a complex, seductively nostalgic picture of the role which the wireless could play in the parlour of the modern home. Again and again they suggested that this newest of new technologies might in fact return the family to a distinctly Victorian fireside, the sphere presided over by a reassuring embodiment of Coventry Patmore's 'The Angel in the House'. She represented a far less alarming image of womanhood than that of the post-war, new-look flapper – one of whose many reprehensible characteristics was perceived to be her restlessness. The first ever Easter edition of the *Radio Times* was overt about this. In an article entitled 'Women and Wireless – Radio a Great Home Maker' the (male) author suggested that broadcasting was bringing to a close the recent and lamentable fashion for women to snatch up their latchkeys and dash outdoors to enjoy themselves beyond the confines of the home:

The *Radio Times* marked the tenth anniversary of the BBC with this full-page cartoon in November 1932, which showed how the technology had changed in so short a time as well as common ideas about men, women and wireless.

> Wireless broadcasting is unconsciously giving us new ideas – or, rather, restoring the ideas of out-of-date conventions. During the past few years, home has had rather a bad time. It has been neglected because it was dull, and because housewives, after the drudgery of the day, have insisted upon taking their pleasure out of doors.[7]

He then quoted a (probably fictional) grandmother saying that wireless is 'a monotony-breaker, a loneliness dispeller and a great homemaker', before suggesting 'Soon we shall be singing: "What Is Home Without a Wireless Set?"'

His appraisal was largely accurate in predicting that most homes would soon own a wireless set and that those that did not would be pitied. Even so, at a time when more professional and employment opportunities were opening up for women outside the home, it reinforced a reactionary message about their traditional place. This is a good moment also to point out a surprising fact about the composer of the parlour song which he misquotes here with humorous intent, and on which others had riffed in cartoons and articles. 'What is Home Without a Mother?' was one of the famous and much-loved mid-nineteenth-century American 'Hawthorne Ballads', which were commonly understood to be the work of the composer and librettist Alice Hawthorne. In fact they were by a man, one Septimus Winner, who used this female pseudonym and a fake persona, which he extensively embroidered over three decades, as a fabulously successful marketing ploy.[8] It's a timely reminder that the story of women listeners in the home should be told as far as possible through the voices of the women themselves.

⁂

'Radio has brought revolutionary change into the lives of millions of women': this was the view of Margaret Bondfield, the Labour MP and women's rights activist, in a 1937 *Radio Times*[9] article about how vastly the new medium had expanded women's lives over the last fifteen years. She went on to say: 'The slender wire brings the world and its affairs into the tiny kitchen and living rooms which hitherto had isolated so many housekeepers in the performance of their duties.'

Illustrating a November 1937 *Radio Times* article, entitled 'What Radio Can Do for Women' by Margaret Bondfield MP, this drawing suggests how the broadcasting schedule encouraged housewives to take a break during their busy days.

Radio Times listings give some idea of what a woman might actually hear of 'the world and its affairs' on a day-to-day basis and provide a starting point for imagining the impact of wireless on her life.

Broadcasting did not begin until 10.15 a.m. (this was the case right up to 1939), so there were several silent hours after husbands had departed for work and children for school. The opening programme was the *Daily Service* (announced in its own copperplate font). This immediately lifted the day for Mrs Privett in Barton Hill: 'I brings the potatoes to the table to peel, to listen to the services. I do. I couldn't hear from the sink. No difference from churchgoing. That half hour doose me good.'[10] There was much discussion about religion on air: was it the same as attending church (the BBC made a point of scheduling its services so that they never clashed with parish timetables), were outside broadcasts from cathedrals and a range of denominational churches across the country preferable to studio transmissions, and – most importantly – did the various ministers achieve the right tone? Before an early retirement caused by poor mental health and acute asthma, the Reverend Dick Shepherd had been universally popular for his friendly voice and the humanity of his message. In Barton Hill, more than one resident suggested that listening to him had helped families become less argumentative, more peaceable: truly a compliment.

Some women complained that any kind of listening was too distracting. In reply to Winifred Gill's question about whether radio helped the daily drudgery, another Barton Hill resident poignantly remarked: 'I dances when I hears the wireless – I couldn't get on with my work.'[11] A Warrington woman later recalled that the radio timetable had not fitted into her relentless routine: 'I never had time. There were too many jobs to do, what with baking and washing and all that. In any case, the radio was in the parlour, whereas I'd be in the back most of the time.'[12] But it was becoming common for women to perform portable jobs – peeling, chopping and cleaning – to the companionable sound of the wireless, to synchronize their own schedules so that they could be in the right room at the right time. A middle-class subscriber to the *Radio Times* described how she would settle with her mending basket and sew to the sound of programmes which she had carefully ringed with a pen in advance, while a domestic servant would pull out the brass stair rods and polish them as she listened to her favourite morning programme.[13] Life as a housewife was busy but it could also be lonely, and radio brought a sense of outside stimulation to the monotonous routine, as another Barton Hill resident revealed: 'The woman in the flat above me only this morning tuned into something she wanted to hear, and then the milkman came. She went down to the door and as she come up, she says to me: "Blast that milkman, now I've missed just what I wanted to 'ear".'[14]

Much of the day was filled with music. On the day of the general election, women could listen to the second part of Verdi's *La Traviata* on gramophone records, as well as a concert of light classical music played live in the London studio. Later, there was an outside broadcast of evensong (copperplate again) from Westminster Abbey. Lighter dance music – the most popular form of all broadcasting, particularly among younger women – was generally kept back for the evening, though on 30 May live election coverage continued well into the small hours.

A dainty tea set, a china basket of biscuits and a top-end wireless set: these elegant women in an advertisement from November 1923 are already enjoying the radio strand specifically aimed at them.

In the earliest days of the BBC there had been a brief experimental slot called *Women's Hour* – two late-afternoon talks running back-to-back and expressly intended for women. The inaugural talks were given by Princess Alice, Duchess of Athlone, on 'The Adoption of Babies', followed by Lady Duff-Gordon, the famous designer (and *Titanic* survivor), on 'Fashions'. This had soon been dropped (*Woman's Hour* did not begin until 1946), but fifteen-minute talks were the absolute building blocks of the daytime schedule. BBC staff returned frequently to a key question: should the topics chosen recognize and reinforce the domestic reality of women's lives or lift them out of it? As early as 1924, one female producer was humorously acknowledging this dilemma in an article in the *Radio Times*: might a talk on a cure for constipation be substituted with one on a tour of Constantinople? Did women want advice about restocking the kitchen cupboard or a celebration of the English countryside? Might debates about diets be replaced with topical questions? Could calories give way to careers and hobbies?[15] Of course, no two women were the same. One of Winifred Gill's interviewees shyly confided that 'anything to do with whales or whale-fishing'[16] spelt romance to her, and she was not alone in this desire to be transported to a world so very different from her own. Mrs Shepherd concurred: 'I likes to hear about other countries, travellers and adventures – and about whaling.'[17] For the broadcasters it was a constant balancing act. For women at home – particularly those who could not stretch to a weekly copy of the *Radio Times* – daily listening must sometimes have seemed startlingly random in content though attractively bite-sized and digestible in scale.

In the election week, women were invited to learn about bridge-playing, adventures with birds, and travels in the French chateaux country; given practical guidance on the washing of blankets and woollens, beekeeping as a profitable hobby, and staying cool in a heatwave; encouraged to partake of Sir Walford Davies's long-running and popular series on the *Foundations of Music*; pick up what they could from the fortnightly bulletin by the Ministry of Agriculture; and gain some cultural updates from James Agate on the theatre, G.A. Atkinson on the cinema and Vita Sackville-West on the latest novels.

By 1928 there were already new series of talks on women's participation outside the home. These included an innovative series entitled *Questions for Women Voters*, which had specifically been conceived to educate them in the rights and responsibilities which would be theirs in the forthcoming election. Simple in format, each episode introduced a speaker who would address an issue such as 'Should Women Be Paid as Much as Men?' and 'Should Married Women Work?' Just how timely such questions were is highlighted by the layout of the

Professor Meek, broadcasting his views on Love and Marriage, enjoys the delicious and altogether unique experience of having his wife so situated that she cannot answer back.

Even the scratchy style of this misogynist 1924 cartoon suggests the depth of marital disharmony between husband and wife.

Radio Times page containing the billing for the episode entitled 'How Laws Are Made'. Directly beneath details of the talk, the editor had seen fit to place a hackneyed cartoon on the old loudspeaker theme: in this one, a small man and dog cower before the ostensible true 'law-maker' of the home – an abusively overbearing woman.[18] In clothes and surroundings, this couple is clearly meant to be working class, but there are other cartoons set in middle-class homes which reveal a similar fear and dislike of powerful women. One, from May 1924, is a split image: on the right, a woman, who is depicted as well-off and educated but careless of her unkempt appearance, stands seething before the amplifying horn of an expensive wireless set. She is listening to her husband, who, in the image on the left, is safely speaking from a studio elsewhere, and intones these words: 'we perceive in this, women's obvious inferiority to man'. The caption reads: 'Professor Meek, broadcasting his views on Love and Marriage, enjoys the delicious and altogether unique experience of having his wife so situated that she cannot answer back.' Of course, these were only cartoons, but they were cartoons appearing in the BBC's official publication, so they give some idea of prevalent levels of casual sexism. For women listening against such a backdrop, just to be made aware of programmes which framed and addressed important questions about their role in society must have felt like a window being opened into a more equal life.

From May 1929 there was a regular new strand consisting of two daily broadcasts entitled *Household Talks*. As the name implied, the focus of these was the domestic sphere. On the Thursday morning of the election there was an episode from a short series entitled *The Growth of Your Child*, in which the Hon. Mrs G. St Aubyn considered 'The Difficult Child'. In the afternoon, the subject of 'Angora Spinning' was under discussion. Other talks during the week included 'Economics in the Home: Why Prices Rise and Fall' and two days of 'Menus and Recipes': 'Plain Loaf Cake' on Tuesday and 'Fruit Creams' on Friday. A little more lightheartedness wouldn't have gone amiss, but it was a good mix.

⁂

The creative powerhouse behind this ambitious output was the BBC's first Director of Talks, one of a very small number of BBC women in creative rather than clerical roles. The redoubtable Hilda Matheson had been appointed by Reith himself in late 1926, and by November 1929 had inaugurated the longest-surviving item in her extensive legacy – *The Week in Parliament,* which runs to this day under the title *The Week in Westminster.*[19] Each Wednesday morning, listeners could hear the presenter, Margaret Bondfield, introducing a female MP (later there were also male MPs), who would give a talk intended quite specifically to educate this first generation of newly enfranchised women about British politics.

THE BRITISH BROADCASTING CORPORATION
HOME, HEALTH AND GARDEN
Price One Shilling

BBC pamphlets such as this one from 1934, which printed recently broadcast versions of *Household Talks*, hint at the optimistic belief held by the first female producers that wireless could improve women's lives.

Like Reith, Matheson was a child of the Scottish manse with a deep moral seriousness and belief in the transformative potential of education, but the two were wildly different in other ways. Fluent in Italian, German and French, educated at the forerunner of St Anne's College, Oxford, Matheson was left-leaning and included among her friends and acquaintances E.M. Forster, H.G. Wells, Rebecca West, George Bernard Shaw, John Maynard Keynes, Virginia Woolf and Vita Sackville-West. In fact, Matheson and Sackville-West had recently embarked on a passionate love affair.

Hers is a story worth relating. This is partly because there is an irresistible counterfactual (alas, only counterfactual!) history to be spun in which, after Reith's departure in 1938, the immensely capable Matheson becomes the first (and, to this day, only) female director general of the BBC. Which, in turn, begs the question, how might output have been different under her culturally broader and socially more open-minded leadership? Mostly, though, it's important for a rather intangible reason: David Hendy has shown uncommon understanding of the frequently overlooked fact that radio programmes are produced by radio producers, and often with a remarkable degree of independence (far more so than in television). For this reason he argues that the personal biographies, passions and beliefs of exceptionally talented individual producers really matter, because these regularly, if semi-anonymously, find their way into the homes and the ears of millions of listeners. Reinforcing the idea, Kate Murphy, *the* pioneering historian of women's working lives within the BBC, points out that this was all the more significant at a time when there were so few women in such a male environment.[20] A passage in one of Matheson's long letters to Vita Sackville-West describes an in tray overflowing with letters from anti-vaccinationists, members of pacifist and women's organizations, promoters of the idea of a channel tunnel, Indian musicians and speakers on infant welfare. Hers was the desk at which decisions about which talks should appear on-air were made, and these were the talks which women heard when they tuned their wirelesses and sat down to listen in their homes. There was also the question of delivery: because Matheson was passionately committed to the novel idea that broadcasters should not speak *at* listeners but make them feel that they were in the room together, talks began to sound rather friendlier in tone. In fact, they were now sometimes just called 'chats', and the intimidating term 'syllabus' which had previously been used to describe forthcoming series was quietly dropped from BBC literature.

By the election year, though, Reith and Matheson were increasingly at odds. At times their disagreements were over petty issues: a series of classic BBC memos finds Matheson wanting to feature recipes for fruit wines, which might economically be made from a glut of garden produce or berries picked free from the hedgerows.[21] Reith forbids this, decreeing that 'no intoxicants should be included'. But their dispute came to a serious head when Reith refused to allow Sackville-West's husband, Harold Nicholson, to name on-air two iconic novels, both banned at the time: *Lady Chatterley's Lover* and *Ulysses*. Nicholson was well known to be left-leaning and to have supported the Welsh miners during the General Strike, but Reith also personally abhorred modern literature. In 1931 Matheson gave up the unequal fight and tendered her resignation.

The mark of Hilda Matheson – and of the four women who worked under her – was profound. The talks they commissioned and produced were entertaining and quietly empowering a generation of women. In February 1929, one listener from Manchester, signing herself simply A.B, recorded her moving response in a letter to the *Radio Times*:

> Such talks come as a god-send to women bursting with mental energy, yet who must stay close to work-a-day household duties. To one, at least, the task of cleaning a kitchen went down a little better whilst listening to the intelligent observations of an intelligent woman.[22]

And in that same year, Matheson, who had generally regarded letters from listeners as a 'stupid and unreliable index', shared with Vita Sackville-West her willingness to make exceptions:

> One very nice one from a working man's wife who hates household work and longs for time to read books and go out to things and can't and she said her ironing and cleaning were polished off with much less crossness because of being able to listen to a sensible woman talking about events of the week.[23]

Over the following years, groups of women interviewed in Barton Hill expressed enthusiastic approval for the regular Friday morning talks by a medical doctor: 'Ought to be more', one suggested, 'I think young mothers is silly not to listen. It'd pay them.' Gardening programmes, too, were considered practical and much praised. The very first garden celebrity had been Marion Cran, who was well liked for her appealingly self-deprecating 'chats', but from the 1930s the famous Mr Middleton became particularly popular: 'We've a beautiful crop of potatoes this year through listening to he. Arran Pilot he recommended, and my chap put them into his allotment – and I've got a great sackful this minute.'[24]

In spite of Mr Middleton supplanting Ms Cran, the daytime talks had definitely begun to make some small inroads into the lamentable on-air gender balance, and women could now hear more female voices even than five years previously. Unexpectedly, this seemed sometimes to make them particularly critical, as though a listener's awareness that she was of the same gender as the broadcaster made her more alert to their class difference. This was particularly true of kitchen topics. In spite of sincere attempts to provide realistic and affordable recipes, listeners were vehemently critical of the twice-weekly cookery talks they heard. One complained bitterly:

> [The] cooking is too messy. Fiddlin' 'ere and fiddling there. 'Take a bit of stuffed parsley' – she's got servants, that's plain. She's a cook to do it for 'er. We've got all the work of the 'ouse to do beside cookin'. We've no time for fiddling 'ere and fiddlin' there.[25]

Many others agreed: 'Well no one can't be bothered to make them things. Too messy, too expensive, too dependent on a large kitchen.'[26]

In the mid-1930s, the exciting new approach which Olive Shapley was pioneering in Manchester was also taken up by one of Matheson's team. For a series in which 'ordinary women' shared their experiences on-air, they were largely identified by their husband's occupations as 'a miner's wife, a policeman's wife, a tailor's wife' and – finally! – 'a countrywoman'.[27] Even so, they must have made lively listening for millions of women in this ever-expanding, specifically female 'imagined community'. The same was true of the sharing of recipes by women who (it was expressly stated) were catering on modest budgets in every corner of the country: Mrs Tarrant in Lancashire on how she made breakfast sausage, Mrs Sutherland in Scotland on her Braemar scones, Mrs Johns on her Welsh queen cake, Miss Tweedy on her Irish soda bread, and an even further-flung and particularly lively contribution from Miss Kate Lovell, who shared the recipe taught to her by an old uncle, formerly an army officer, for an Indian curry.[28]

This sort of output was not radical broadcasting. Its intent was to brighten and improve in small ways the daytime hours of women living within a prescribed status quo. Series titles such as 'How I Keep House' and 'How I Manage' made this clear, but very few women felt that their own days spent keeping house and managing were not brightened when they switched on the wireless.

*

Of course, it was not only married women who took to radio. In Bristol, Miss Vile spoke of the companionship and sense of security listening could bring: 'a friend of mine who is very nervous when alone doesn't mind when she can listen to the wireless'.[29] This was a common sentiment, extending particularly to elderly women and those who were housebound due to physical disability or impaired sight. Another woman shared the observation that 'It's company for people. I know a woman who's lost her husband and she was all alone and she says with the wireless on she never feels lonely. It's like having someone else in the room.'[30] Mrs Cunningham's view was even more utopian: 'I shouldn't think there can be any lonely people left now. The wireless makes company for you at once.'[31]

For the growing numbers of what were sometimes known as 'bachelor girls' – unmarried working women living alone in bedsits – radio provided a boon in the evenings. The same was true for women in domestic service, still the largest occupation outside the home. The end of the war had brought about a deterioration in what the newspapers persistently dubbed 'the servant problem', as girls and women recognized that the sort of employment they had been urged into, in mills and munitions factories, in order to contribute to the war effort, might in fact be preferable to lives of invisible drudgery and loneliness in peacetime service. The (slight) leverage this gave them might even extend to a demand for a wireless. In April 1925 Virginia Woolf wrote in her diary that, if her books were a success, she hoped to fix up wireless for her servant Nelly at Monk's House in Sussex.[32] First-rate cooks were considered particularly hard to find and particularly easy to lose, a situation which upper- and middle-class writers often lamented. E.M. Delafield, for example, wrote in her peerless *Provincial Lady* fictional diaries about her perpetual fear of offending her cook, lest she hand in her notice:

> Cook sends in a message to say that there has been a misfortune with the chops, and shall she make do with a tin of sardines? ... Mem: Enquire into nature of misfortune in the morning. Second, and more straightforward Mem: Try not to lie awake cold with apprehension at having to make this enquiry.[33]

A *Radio Times* cartoon – far from the only one of its kind – depicted a middle-aged buxom woman in a cumbersome black coat and bedraggled fox-fur stole (the regular image of a cook and very similar to the one described by Delafield). She is disdainfully turning down a position, having discovered that the kitchen is equipped only with a small crystal set and not the superior sort of valve radio to which she now considers herself entitled.[34]

Some of the commonest early technical problems were fading. As mains electricity was very gradually installed, even working-class homes became cosier and housework a little less arduous. A radio was often the first electrical gadget to be purchased, and by 1932 mains sets were sold in greater numbers than battery-powered ones,[35] which meant that headphone-listening was becoming a thing of the past. When children returned from school and husbands from work, listening could now be communal, which sometimes proved a source of shared pleasure and lively debate, sometimes of dangerous tension. The Barton Hill grocer Mr James was adamant that radio had had a huge benefit on the home and that women were both the creators and beneficiaries of this change: 'Yes, the mother has come into her own matronly position in the family. One comes in and she gets tea for them, and then another, and at last she'll say: "now clear the cloth, there's a good play coming on". She's got it all arranged, you see, all thought out.'[36] Some women concurred that their control of the wireless switch reinforced their status – 'My 'usband leaves it to me'[37] – while a man described the amicable arrangement in his household: 'My wife, she loves plays. She says: "you let me 'ave the play, and after that you can 'ave what you like".'[38] Others wanted conversation at the end of the day and felt disappointed or resentful when their husbands came home and just wanted to listen to the wireless. There was also a resigned acceptance that male control was the norm, with radio sometimes an ugly new focus for gender wars. Mrs Hennessy commented of her husband: 'He always listens to the 6 o'clock news. You daren't speak... Now Mr Middleton, you've got to clear out for him. You mustn't breathe.'[39] Recollections in Warrington confirmed a similar tyranny, both from a wife:

> I had to sit with my arms folded while he was fiddling with his crystal. If you even moved, he'd be going 'shush, shush', you know. You couldn't even go and peel potatoes, because he used to say he could hear the sound of the droppings in the sink above what was coming through the headphones.[40]

And from more than one daughter:

> My father, he was a bit short-tempered, and he'd be saying, 'Would you bloody well shut up' – threatening us if we opened our mouths, if we made any little noise. Oh God, we daren't move when my father had that wireless on. None of us dared move a muscle.[41]

For another young Warrington woman, though, radio was a boon because it reduced her sense of what her twenty-first-century peers would call FOMO or fear of missing out, by bringing into her home the dance-band music that her friends enjoyed in venues which her own very strict father, a policeman, considered 'dens of iniquity'.[42]

Many women agreed that, with a little give and take, the radio could enliven an evening, and indeed bring the family together in the home (a subject covered in the next chapter), but others expressed a sad disinclination to argue about the content of what they were hearing: 'No ... not worth it. Not worth starting an argument over something he knows something about and I don't.'[43] Casual references to domestic violence pervade the popular culture of this era, though few are as chilling as this real-life instance of coercive control which

Winifred Gill faithfully transcribed in her notebook but – significantly – did not include in the final publication:

> [A] man and wife I know. She had a real grievance against him, he did not give her a fair share of his wages to keep house on. But if ever she started to talk about it, he put the wireless on so loud it drowned her. And when he went out, he'd say – 'now I've tuned in to a foreign station, and I shall expect to find it there when I come in – so don't you touch it.' So she couldn't have it on, you see, while he's away. But one weekend he went away to a conference, and as soon as he was gone, she said to her boy and girl – 'now we'll have the wireless,' but she found he had disconnected it. She's left him now.[44]

By 1938, at least one radio manufacturer was promoting the idea that households might splash out on a second wireless set. The EKCO promotions department sent out letters to radio shops adorned with a bright yellow stamp, jazzy and surprisingly twenty-first century in its image of a woman and man sitting companionably in the same room but back-to-back, each listening on headphones to their own gadget.[45] In real homes, though, statistics speak louder than advertising copy: an equally stylish 1930s' infographic in Winfred Gill's pamphlet makes clear that fathers had more than twice as much say as mothers in terms of who chose the programmes.[46] Sadly, radio was not altering the traditional gender balance that still prevailed in most homes. In spite of his idealized picture of the mother 'coming into her own matronly position', even the ebullient Mr James, probably unintentionally, acknowledged the truth of this: 'I'm gifted, if you can call it gifted, with a dominant personality, and wherever I am, if I want to listen, everyone has to be quiet.'[47]

⁂

On 24 April 1936 the BBC held a conference in London about the Household Talks strand, which was attended by representatives of almost four hundred women's groups. It was extensively covered in the press on the following day, and the *Daily Mirror* reported that 'the most popular speaker was a woman who said that men should be taught by the radio to help in the home.'[48] The fulfilment of such a manifesto was far off indeed. The 'revolutionary change' which, in the following year, Margaret Bondfield attributed to wireless in her *Radio Times* article, was also something of an overstatement. Perhaps more accurate assessments came from women themselves. Mrs Ettery in Barton Hill stated modestly that there was now in women's lives 'much more to talk about. There are so many interesting things (on the wireless)'.[49] A working-class housewife from the London suburb of Cricklewood, who was an early participant in the Mass Observation social research project, wrote more expansively:

> The radio is my inspiration and relaxation. Talks inspire me to deeper thinking and reading, while I often do a tiresome job of sewing while listening. Music makes me caper about madly if I am alone and have excess energy, and to switch on when I am tired and hear something familiar like Hangles Largo [*sic*] is heaven itself.[50]

Such broadcasting might not have been truly radical, but many acorns grow into saplings and eventually into mature oaks, and Hilda Matheson and her team certainly scattered these on fertile ground. The 1929 general election returned fourteen women MPs, an increase of ten.[51] At the same time, wireless voices were increasingly addressing listeners not as the Victorian women their mothers and grandmothers had been, women whose lives the prevailing culture seemed sometimes to want them to replicate; not as shallow, immature flappers, as the newspapers dismissively termed that first enfranchised cohort; but as intelligent, resourceful mothers, home-makers, modern women and active citizens in an increasingly complex twentieth-century world.

A stamp added to a 1938 advertising letter sent to radio supply shops by EKCO contains a surprisingly contemporary message.

RADIO TIMES
CHRISTMAS NUMBER

6d

SEVEN

ONLY VOICES OUT OF THE AIR

> You often hear folks say, 'I must be home by 9, there's a good programme on then.' My husband said to me the other day: 'we shan't have many more Saturday nights out, now that winter is coming on. There's generally a good programme of a Saturday night.' We don't bother to go out – just settle down by the fire.
>
> Mrs Keep, Bristol housewife[1]

By 1932 King George V was no longer a broadcasting novice, but his brief address on Christmas afternoon, from his residence at Sandringham to homes across the country, was a landmark for listeners. As with all radio broadcasts, there had been a good deal of behind-the-microphone teamwork. John Reith had unsuccessfully been wooing the king for almost a decade, but a tactful nudge from the prime minster, Ramsay MacDonald, had at last achieved the desired result. By royal request, Rudyard Kipling had been approached to write the speech. On Christmas Eve the BBC's director of outside broadcasts had travelled up to Norfolk for a final reconnoitre, and on the day itself there were sound engineers at hand to ensure that the broadcast went without a hitch. The king began with these words:

> Through one of the marvels of modern science I am enabled this Christmas day to speak to all my peoples throughout the Empire.

Unmistakably the work of Edward Ardizzone, this captivating *Radio Times* cover from December 1932 brings together key images of a traditional British Christmas while tacitly placing radio listening at its heart.

> I take it as a good omen that wireless should have reached its present perfection at a time when the Empire has been linked in closer union, for it offers us immense possibilities to make that union closer still.[2]

This was a massive shift in scale and ambition from the BBC's earliest Christmas offerings, just under a decade ago. Back in 1923, despite its iconic cover, the first Christmas *Radio Times* had not even mentioned the words 'Christmas Day' on the banner of its listings page for 'Wireless Programme – Tuesday (Dec 25th)'.

Such listeners as there were had had to wait until 6.30 in the evening, when 2LO had at last opened up with the not entirely enticing announcement that the Reverend J.A. Mayo would 'talk to the children', and there would then be a 'Christmas Play' on *The Children's Hour*. The main event had been a 'Christmas Night Programme' featuring comedy favourites Helena Millais and John Henry, and music from the London Wireless Orchestra. At that time – before the unifying reach of the Daventry transmitter, before the BBC's late-1920s' reorganization away from national and local output to a more centralized national and regional system – listeners in Birmingham, Bournemouth, Cardiff, Newcastle and Manchester had all been able to tune in to simultaneous broadcasts from London, but Bournemouth listeners could also hear their own 'Afternoon of Carols', a thirty-minute edition of *Women's Hour* and an almost two-hour-long 'Kiddies' Hour' special, featuring 'De Vekey's Juvenile Serenaders'; while at teatime in Manchester there had been a chance to enjoy a special recitation of 'A Christmas Carol (Chas. Dickens)'.

Just as Dickens had invented the cherished image of a British Christmas back in 1843, so now, in 1932, the BBC was reinventing the day as the quintessential festival of family and home. This time, though, each and every family and home across the country could be linked together in a wider wireless network. Turn the pages of the 1932 *Radio Times* and the offering is impressively lavish and coherent. The full day of broadcasting included all the now familiar features, though a tellingly parochial mid-evening billing stated: 'If there is any News, it will be broadcast at 9.0 pm.' There was incidental music from the theatre and a recital of Christmas songs. The announcement

The Christmas message broadcast by George V in 1932 suggested that nation and empire were bound into a single family.

of 'Part VIII of a series of Talks on English Religious Poetry of the Seventeenth Century' was perhaps the cue for many listeners to nip from the parlour to the kitchen and baste the bird, but all were positively encouraged to be close to their wireless sets for the early evening religious service, and to join in singing the main carol, 'In Dulce Jubilo', for which all four verses were helpfully printed in full. The Christmas Appeal – this year on behalf of the British Wirelesses for the Blind Fund – was given by no less a contributor than the prime minister himself. Without a doubt, though, the highlight of the day was at two o'clock in the afternoon. 'All the World Over' was billed as 'A Message to the Empire', which included live relays from the Irish Free State, an Atlantic liner in mid-ocean, Canada, New Zealand, Australia, Port Said, South Africa and Gibraltar. Then, just after three o'clock, a linking voice back in London had introduced the king as though he were ushering him personally into a room.

Sandringham, with its fussy Jacobean-style architecture and eighty-six fireplaces, was more than just another royal residence – it was known to be the place which the king personally loved best, and he sounded relaxed as he concluded his short address with these words:

> I speak now from my home and from my heart to you all, to men and women so cut off by the snows, the deserts or the sea that only voices out of the air can reach them. To those cut off from fuller life by blindness, sickness or infirmity and to those who are celebrating this day with their children and their grandchildren – to all and each I wish a Happy Christmas. God bless you.[3]

Kipling's words neatly encompassed what was becoming the intertwined mission of the monarchy and the BBC during the early years of the 1930s – to draw together a diverse nation of 46 million people into a single virtual entity. 'Voices out of the air' could now reach the furthest flung corners of the Empire, but they were also at work at a domestic level, attracting listeners back into their own homes whilst simultaneously giving them a stronger sense of belonging to the world outside their front doors. Less than half the population had access to radio, but everyone knew that it was here to stay. A journalist on *The Sketch* described in 1927 how it had embedded itself into daily life

This undated advertisement for the popular Murphy brand plays on the idea that equipment needed to be upgraded regularly. Some retailers would in fact accept old sets as part-payment for newer models.

as if 'a routine like shaving'.[4] Newspapers and magazines were full of long features and short, snappy articles which both reflected the uptake of wireless and proselytized for it. An anecdote in *The Tatler*, for example, related how a villager in the West Country, having come into some money, decides to set himself up as a fishmonger. Hoping to build up his customers, he is rebuffed by the local clergyman, who explains that he gets his fish fresh from town. In turn, the man ceases to attend the local parish church, and when challenged by the vicar explains: 'I bought a wireless set. Now I get my sermons fresh from town!'[5] Of course, the story is probably fictional, but it gives a sense of how investing in a wireless set might both enhance home life and, at the same time, expand the individual's sense of their place in a wider world.

As the listening figures grew, though, the challenge for the BBC was how to satisfy the audience in accordance with Reith's commitment to a single service for the entire nation. The same *Sphere* journalist summed up their diversity in terms of the different time of day that they ate their evening meal (a common determiner of class in British surveys), their interest in classical music or jazz, their preference for debate or instruction, their appreciation of broadcast church services or their belief that these were sacrilege, their opinion that repeated weather reports and football results were intolerable or the very lifeblood of conversation.[6] The holy grail was output of common interest to all.

⇸⇷

The 1932 royal message received one of the heaviest and most appreciative postbags that the BBC – now a corporation – had so far recorded. It immediately became a fixture, the culmination of what was an increasingly busy annual calendar. From the start, both radio manufacturers and the broadcaster had been aware that they needed to push their wares harder during the summer, but growing sports coverage was beginning to give wireless listening year-round appeal, and soon there wasn't a month without some event, match, anniversary or special occasion. Most had existed already, but the BBC had cleverly gathered the dates together and written them up, as if on a notional family calendar hanging in the national kitchen.

John Reith had unilaterally decided that the religion of the BBC, like the religion of the monarchy and the country as a whole, was Christian (and largely the Anglican version of Christian), so Advent and Christmas, Holy Week, Easter and Ascension Day provided a good start. Then there were bank holidays, the feast days of the patron saints of the four nations – St George and St David, St Andrew and, more problematically, St Patrick. An increasingly ambitious outside broadcast team was soon providing coverage of the Grand National, the Oxford and Cambridge Boat Race, the Derby, the FA Cup Final from Wembley, the TT motorcycle races from the Isle of Man, cricket from Lords and the Oval, tennis from Wimbledon. There were also state occasions such as Trooping the Colour and the Lord Mayor's Banquet, as well as cultural events such as Burns Night. The celebration of Shakespeare's birthday on 23 April was a particularly grand affair in 1932 as it marked the moment when the newly designed Shakespeare Memorial Theatre in Stratford-upon-Avon rose from the ashes of a terrible fire which had gutted the old building back in 1926.

Even the wealthiest, most mobile and sports-crazy member of society could never have attended all these events in person, but now

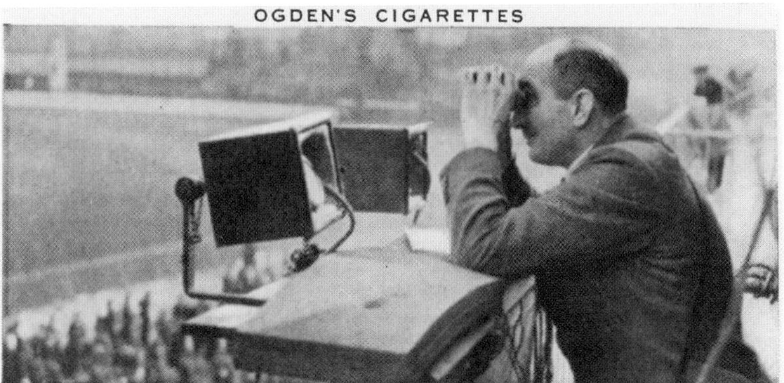

OGDEN'S CIGARETTES
BROADCASTING THE BOAT RACE—THE B.B.C. LAUNCH "MAGICIAN"

Wireless provided many popular themes for cigarette cards, but these broadcasters' views of various major sports events are outstanding in their conception and quality.

they were open to everyone. E.M. Delafield's fictional diarist often remarks on how a dull evening spent sitting by the fire in her upper-middle-class Devon home is brightened by listening to the wireless, but she also has some freedom to travel and experience events in person. In spite of frequent anguished correspondence with her bank manager, she makes trips to Plymouth and occasionally to parties, theatres and galleries up in London. How much greater, then, must the pleasure have been for the hugely expanding number of smaller families living in the newly built suburbs, or for those now marginally better-off people whose status was the result of the social changes observed by George Orwell: 'the sharp distinctions of the older kind of town, with its slums and mansions, or of the country, with its manor-houses and squalid cottages, no longer exist.'[7]

The largely silent voices of listeners from modest homes can be heard in the letter of a quietly courageous man from Birmingham, who signed himself 'G.M.C.' when he wrote to the *Radio Times* on 20 January 1928. Such correspondence always needs reading with a little caution, but this letter provides a moving account of the transformative power of radio. G.M.C. started with a description of the daily grind:

> Many of your readers must be office workers. They must know what sort of a life is that of a clerk in a provincial city – a tram-ride to the office, lunch in a tea-shop or saloon bar, a tram-ride home. You daren't spend much on amusements – the pictures and that – because you've got your holidays to think of. We have no Trade Unions and we don't grumble, but it's not an easy life. Please don't think I'm complaining. I'm only writing to say how much wireless means to me and thousands of the same sort. It's a real magic carpet. Before it was a fortnight at Rhyl, and that was all the travelling I did that wasn't on a tram.

He then went on to evoke what radio meant to him personally:

> Now I hear the Boat Race and the Derby, and the opening of Menin Gate [the memorial to the missing of the First World War, which had been unveiled at Ypres in July 1927]. There are football matches some Saturdays and talks by famous men and women who have travelled and can tell us about places. ... But I do like best the running commentaries. You can just see the crowds at the Boat Race, the football and the boxing

> matches. I don't believe that when you're actually there you realise them in the same way. It is really better than being there almost. I could simply sit and listen to the sounds without bothering what the announcer says. I've seen boxing matches on the pictures but they weren't ever as real as that Albert Hall march. You could smell the cigars.[8]

In light of such enthusiastic uptake, constant advertising was encouraging better-off households to upgrade their sets or to purchase a combination radiogram. The newspapers continued to feature novelty apparatus: northern sweetheart Gracie Fields, for example, with her hair arranged in 'earphone' side buns, endorsed a wireless hidden inside a quartet of antique books.[9] There were also many portable versions on offer. These were nothing like the pocket-sized transistor radios which later revolutionized listening on the go, but fashionable for picnics, motor trips and boat rides. For the sum of a shilling, it was even possible to hire headphones and listen in while travelling north on an LNER train.

But the idea was now growing across classes that staying in might actually have advantages over outside forms of popular entertainment such as a trip to the pictures: during the worst years of the slump, the appeal of not having to wash or get dressed to listen in became stronger, while the annual licence fee or cost of replacing the wet batteries (which non-mains sets still required) could be eased by joining a savings club. One man wrote to the *Daily Mirror* declaring that he had been able to enjoy the Derby better on his cabinet-top wireless than at Epsom itself, where he understood it was hard to find a good viewing-point;[10] a Warrington woman later recalled her father bringing in his drink from the pub so that he could enjoy it to the sound of the wireless ('A lot of people were right homebirds in them days'[11]); while many in Bristol spoke casually about what was, in fact, a major cultural shift: 'Oh yes, people often say they must get home because there's a good programme on the wireless.'[12]

⁂

It's true that once the novelty of ownership had worn off, radio might simply have become a background noise in some homes. Mrs Bennett's remark is a particular favourite of mine: 'My son in law's father, when he comes in of an evening, he switches on the wireless and sits down beside it and goes to sleep. And nothing won't wake him unless they turn the wireless off. "E will 'ave it on"'[13] (not really what any radio producer – then or now – wants to hear). But in the majority of houses radio seems to have brought families together and, in turn, generated conversation. Sandringham might boast eighty-six fireplaces, but in Sandringham Street in Belfast or Sandringham Terrace in Sunderland or Sandringham Drive in Bangor houses were far more likely to have just one, which meant that listening was necessarily a shared family event. 'There's some talks people wouldn't miss for anything. It brings people together as a family'[14] stated one Bristol woman, while another frankly recalled how limited the topics for conversation had often been in the past:

> People don't quarrel so much in the home. They've more to talk about. Their troubles is 'alf forgot. Why in the old days, if there was somebody dying or somebody bad, it was all the talk. You 'eard nothing else. Nowadays you do just 'ear it, but that's all.[15]

Family bickering might be replaced by more constructive types of argument. Mr Beedell declared fulsomely, 'the man as invented the wireless, invented a wonderful thing', before continuing:

> People 'as learned more, and there's more conversation in the home. You get the family sat in the house of a night, and there's a talk on the wireless, and someone doesn't agree and pulls it to pieces. Then they all 'as a go and gets outside of it. I've known 'em argue for hours. It's bound to make more talk in the home.[16]

A family might sit up late to listen to the end of a play, sometimes in the dark, by the flickering light of the fire. Live night-time relays of boxing matches became increasingly popular in transporting people away from their homes. A journalist for the *Hartlepool Northern Daily Mail* vividly reported the impact of the big Tunney–Heaney fight, when 'a succession of thrills with interludes of sparkling humour flashed across the ether in full-blooded American accents'.[17] Until 1927 there had been an embargo on sports broadcasting because of pressure

miles and "keeping on keeping on" for several hours, with a point of half-a-dozen miles thrown in.

Thursday was not too good, a Silk Wood fox surrendering at Bowldown, and hill-hunts filling the bill later from Boxwell, and then the Kilcot end of the range. Whither "Beaufortshire"? Mostly to Aintree, not Lansdown, on Friday! Master amused himself with the ups and

MRS. F. H. BREWSTER and MRS. W. E. H. GARNER took a portable wireless set when "hunting by car" with the Quorn when hounds met at Thorpe Satchville, so that they could listen-in to the broadcast of the Grand National while attending the meet!

A portable wireless apparently provided these two women and their chauffeur with the experience of simultaneously listening in to the Grand National while following the local hunt in April 1934.

from the newspapers, but regular match coverage proved surprisingly popular. Mr James, the grocer, grew to love it: 'Confront me with a game of tennis and I shouldn't know what it was about … or boxing … there was folk who ordinarily would be shocked at flat noses and thick ears, came into the shop and talked with abandon.'[18] Radio was also making sport less of a male domain: 'I don't believe there's a woman took an interest in football before we had the wireless – not without she was proper sporty. Now I always listen to the commentary, though I wouldn't want to go to a match.'[19] Live coverage of Test matches was welcomed by men who could not afford to take three days off to spectate in person but might have the wireless on in the background: 'During Australia's first innings, I said to my wife "If they bowl Bradman for less than 14, then that pound note on the table is yours, my dear", but they didn't.'[20] There were real novelties, such as motor racing and ice hockey, as well as a sense of involvement in sporting events which had previously, almost by definition, only been available to certain classes: 'look at the Boat Race, 'ow that's

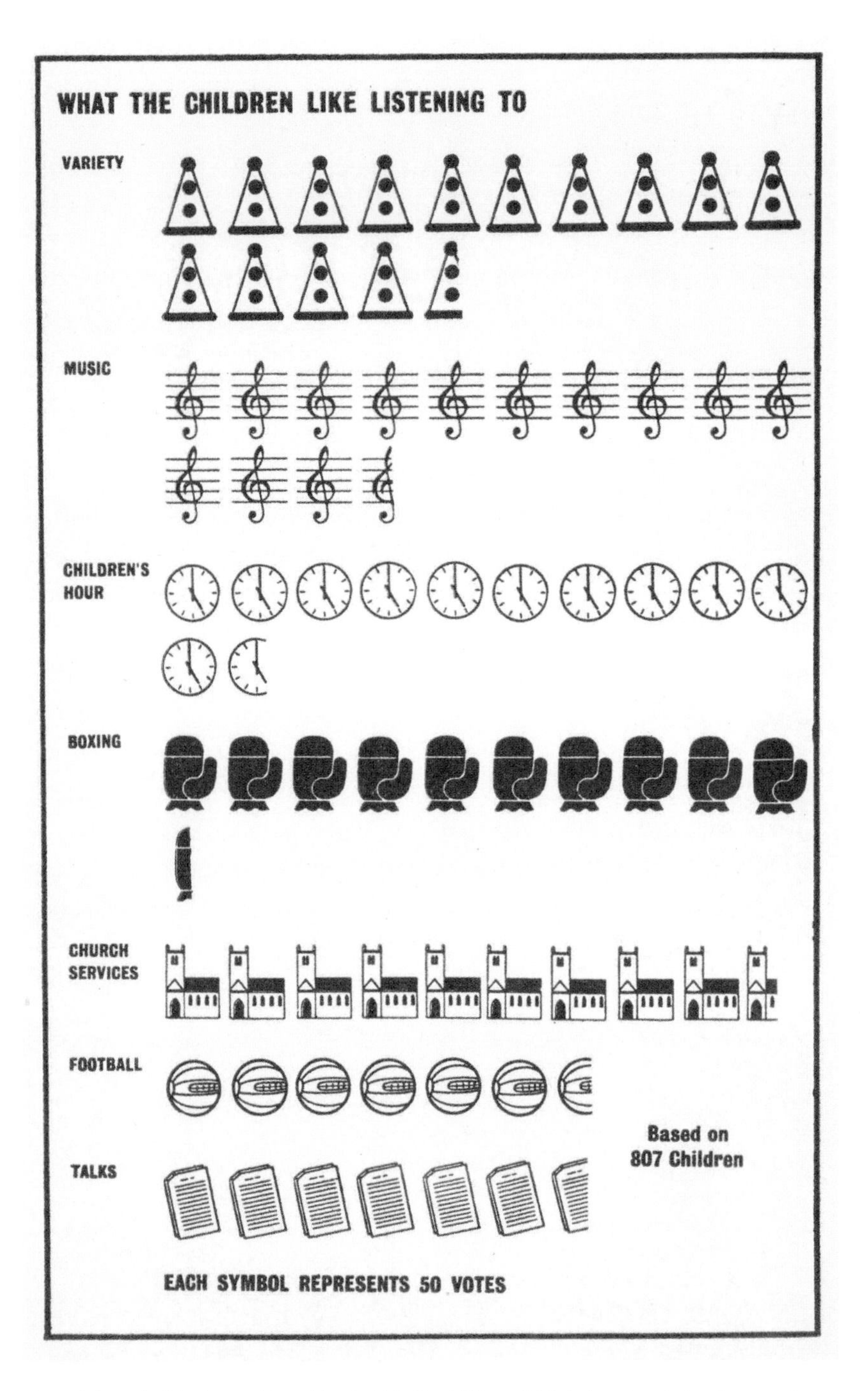

One of three bold infographic tables in Winifred Gill and Hilda Jennings's book, revealing the listening preferences of children in Barton Hill.

come to the fore. There's more sports enthusiasts now than ever. It builds up the spirit, and that brings on other things.'[21]

Children often shared in whatever their parents were hearing. In Barton Hill many were allowed to stay up for the boxing matches or sent to bed at the normal time and then brought downstairs later. Filling in her questionnaire, one small girl wrote: 'I goes to bed and wakes at the right time', while some children were actually put in charge of waking the family: 'Our Jack banged on our bedroom door. "Wake up, Dad," he said, "Time for the Fight," but before my husband was properly roused, it was all over.' There's just a hint here that radio was once again bringing about a shift in authority between the older and younger generations, with a small child temporarily in charge. For others, though, it could prove too much: 'We all sat on the stairs and we heard it all, that is, except our Tommy. He fell asleep in the middle and lost half of it.'[22] Such intense experiences of outside entertainment were truly a novelty, but some mothers were concerned that shared listening was exposing their children to frightening dramas and distressing real-life news stories. This was particularly the case in working-class homes where space was tighter and bedtime traditionally later.

Technological change is, of course, traditionally accompanied by moral panic about its impact on children, and wireless was no exception. In the *Liverpool Echo* one headmaster shared his view that

> wireless is one more difficulty in the way of getting homework properly done. We have always had to contend with distractions. Boys very rarely give their attention to work done at home. If it isn't a loudspeaker … it is a little sister practising scales, or someone else doing the loudspeaking.[23]

A headmistress expressed fears which resonate (albeit about social media) today: 'The danger [arises] when school children [try] to divide their attention between the steady, carefully prepared and progressive system of school teaching and the desultory interests of the radio.' Parents should be firmer, she suggested, and 'more self-sacrificing of their own listening habits in removing the hindrances to the obtaining of a disciplined mind'.[24] Interestingly, she then added this rather intriguing observation: 'to use the appropriate terms, I might almost say that the most pitiful condition anyone can be in is to have a broadcast mind.' There were also concerns that the democratic nature of wireless might somehow contaminate the minds of more privileged children. The headmaster of Rugby School revealingly shared his fears with *Daily Telegraph* readers that 'people would listen in to what was said to millions of people, which could not be the best things',[25] and a correspondent to the *Evening Chronicle* worried that wireless would produce 'all-alike girls'.[26] Other head teachers were far more positive, commending the practical experience of constructing a set (a popular hobby throughout this period) and the beneficial content of some programmes.

Inevitably, there were local casualties – a school sports day poorly attended because it coincided with a broadcast of the Derby, a sense that windows were less frequently left open than in the pre-wireless past, or that there wasn't quite so much neighbourly chat, but such objections were outweighed by common agreement that wireless was enriching home life.

Sports coverage, talks and plays were providing something entirely new to domestic life, but the fresh source of music might risk supplanting the well-established homegrown variety. If the BBC was fast becoming the key provider of cultural life to the entire nation, the piano in the parlour seemed increasingly to represent home entertainment from the era before the coming of wireless. Some understandably expressed nostalgia for a time when courting couples had drawn up round the fire to join in a spontaneous chorus of popular songs on a Saturday evening, or the family had gathered for a shared rendition of 'Abide with Me' on a Sunday night. Many agreed with the sentiment that 'a song in the room is worth two on the wireless' and pointed out that wireless broadcasts did not necessarily coincide with a desire for music in any individual home, but they also confided their view that it was easy to overstate the pleasure of being invited to listen to somebody else's child bashing out 'The Bluebells of Scotland'. And if one mother acknowledged that 'Edna did learn and I think she'd have persevered, only it wasn't worthwhile, you see, when you can get it on the wireless and no exertion',[27] others, including piano teachers, sellers of sheet music and of a wider range of instruments, rather agreed with the ever-expansive Mr James, who was of the decided view that the wireless was in fact infusing domestic music-making with a fresh lease of life:

> There's people sometimes from my own county of Monmouth and you'd wonder how they'd come to be almost proficient... My father was an expert euphonium player, the only euphonium player in the band. And yet he wasn't so good as people are nowadays. The wireless encourages proficiency. Now there's a boy next door plays the mouth organ. I say to him, 'you keep on with it, you persevere, and we'll be hearing you on the air one of these days'.[28]

Although the piano had traditionally been a female instrument, Mr Corfield pointed out that boys now heard 'the piano played so many different ways on the wireless, they takes an interest. Then the band leader is generally a pianist. They fancy they'd like to get up to the standard of Charlie Coombs.'[29] One canny manufacturer came up with the idea of a combined upright piano and wireless, with the receiver and speaker neatly placed in a panel where the music stand would normally go, and the iron frame acting as an aerial.[30] This made it possible – at least in theory – to play along to music on the wireless. Entirely British-made, it was irresistible – apart from the cost: 52 guineas would eat up two-thirds of the annual pay of a factory worker or agricultural labourer, and two months even of a teacher's salary.

Piano sales in the 1930s were dropping sharply, with a newsreel item featuring the poignant story of a bankrupt manufacturer burning all his unsold stock on a vast bonfire. Magazines such as the *Radio Times* carried humorous articles lamenting the possibility that the family sing-song might now be a thing of the past. The 'piano wars' were not yet quite over, but there was no stopping this shift in the symbolic value of the instrument. Soon, a new BBC publication, *The*

Listener, carried a version of the long-running advertising series 'You Can Be Sure of Shell', in which a split image captures the contrast between then and now: on the left a corseted Victorian lady is histrionically performing an 'Air on the Piano', while on the right her stylish modern counterpart smokes a cigarette as she casually switches on a sleekly designed wireless set to hear the 'Piano on the Air'.[31]

⁂

Wireless broadcasting could also produce less predictable effects. Unlike the experience of watching an emotional film in the anonymous darkness of the pictures, a powerful programme transmitted into the home might arouse reactions which could be uncomfortable for family members living in close proximity to one another. The great Irish playwright Brian Friel, who was born in 1929, brilliantly lays bare this power in *Dancing at Lughnasa*, which is set in the years immediately after the new high-power transmitter at Athlone had made radio available to most Irish people (rather later than elsewhere), in 1932. The play opens with the adult Michael recalling his boyhood: 'When I cast my mind back to that summer of 1936 different kinds of memories offer themselves to me. We got our first wireless set that summer – well, a sort of set; and it obsessed us.'[32] Loosely based on the lives of his mother and four unmarried aunts, who shared a low brick cottage in a small town in the south-west of County Donegal, their erratic wireless set, nicknamed 'Marconi', has pride of place in the kitchen. While both the charming Welsh wastrel Gerry Evans and their own disgraced brother Father Jack bring disruption into the sisters' precarious lives, it is Marconi which truly releases their passion during the late-summer pagan festival of Lugh. Marconi's intermittent music – its battery is constantly running low, its aerial unpredictable – drives the sisters almost wild, fanning their desperation and desires. During the most telling scene, when the sisters dance frantically, Kate, the oldest and a dignified schoolteacher, has to take herself out into the garden to express fully the feelings coursing through her body. In the stage directions, Friel describes the moment when the music comes to an abrupt halt, mid-phrase:

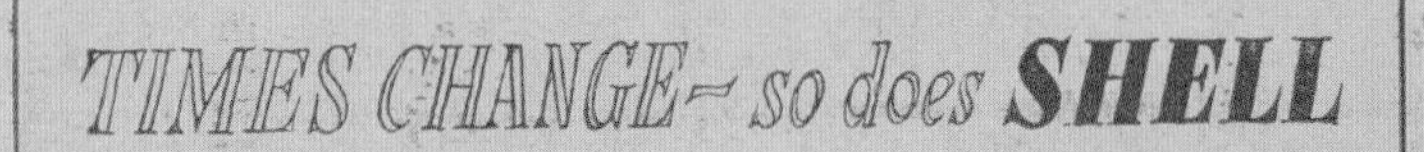

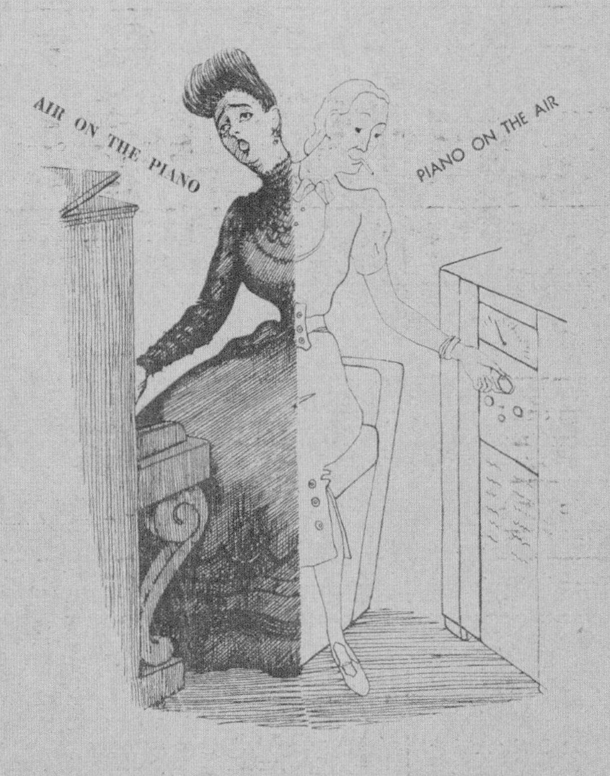

You could not switch on the piano; but it is just as easy to turn on the radio as it is to start your car. When you are speeding along a by-pass think for a moment of the difference between motoring thirty years ago and now. The change is due to the steady development of the motor engine towards an even higher standard of efficiency. Shell, too, is constantly aspiring to new and higher standards and it is this policy of gradual change that maintains its high reputation

YOU CAN BE SURE OF SHELL

Printed in England by W. Speaight & Sons, Ltd., 98-99, Fetter Lane, E.C.4, and Published by the British Broadcasting Corporation, 35, Marylebone High Street, London, W.1, England.—All Editorial communications to The Editor, The Listener, Broadcasting House, London, W.1.—19 *January*, 1939

The advertising catchphrase 'You can be sure of Shell' was only a couple of years old when it appeared in this 1939 version, capitalizing on the by now familiar idea that radio had transformed most homes.

> Silence. For some time they stand where they have stopped. There is no sound but their gasping for breath and short bursts of static from the radio. They look at each other obliquely: avoid looking at each other; half smile in embarrassment; feel and look slightly ashamed and slightly defiant.[33]

In Bristol, a number of women confessed to similarly mixed emotions, one commenting that the boxing made her heart beat faster, another that it seemed to upset her nerves. Most vividly, one confessed: 'I feel dreadfully embarrassed sometimes if a play is on and it gets very sentimental or very passionate. I don't mind so much if its Joyce with me, but if I'm here with Les (son aged 17) I feel ever so uncomfortable.'[34]

As the wireless calendar grew, so did its twofold impact. It spoke to listeners individually and made life at home more attractive, while collectively gathering them – or attempting to gather them – outwardly into a stronger sense of a single national identity. The scale of this impact might be as slight as a tacit agreement that it was probably best not to hold a choir practice on an evening when Gracie Fields was on-air ('I think she fetches us all home'[35]), or as great (if intangible) as an increased sense of inclusion. One man suggested that radio was now giving 'people a chance to hear about other parts of the world, feel themselves a part of things – make them feel more important, not left out of things'.[36]

From the beginning John Reith had been a passionate believer in the unifying potential of radio, and the 1932 royal Christmas broadcast aligned the BBC's mission with that of the king, who was, in the following few years, ever more clearly to position himself as the head of a large family, diverse in background and interest but unified by their common nationality. It helped that the majority of the population identified themselves as Christian and only a tiny number as republican, but Great Britain was made up of England, Scotland, Wales and Ulster, the speakers of at least four languages. A listener from any of these nations might feel under-represented, even when what they heard had originated from close to home or was apparently aimed specifically at them. An English-speaking Welsh woman, for example, listening to her wireless in a big industrial city in the south, might feel far distant from a Welsh speaker in a rural region of the north. Language was a far thornier issue in a nation where over 37 per cent of the population spoke Welsh than in Scotland, where Gaelic was spoken by only a small minority. Listeners from other groups must have felt even less represented – the tiny numbers of Jewish people, members of the Chinese and African communities based in port cities, Eastern Europeans and Americans.

Celebrating diversity wasn't hard to pull off at a broad-brush, symbolic level, as the schedule for St Andrew's Day in 1932 illustrates. The evening featured a 'Wireless Pageant of the People of Scotland' (apparently all male) titled 'Hail Caledonia'. In this an old gentleman invites a range of Scotsmen to his home in St Andrews to engage in a playful competition: the Celt speaks in favour of the mountains, the borderer of his southern hills, the man from the north-east of his own ancient corner of Scotland, and the Ayrshire man of the land of Robert Burns. All finally agree that wherever they come from – and this includes those who are south of the border or across the seas – they are united in being Scottish. Well intentioned, such celebrations were inevitably reductive. A programme similar in intent was broadcast from Cardiff on St David's Day, but St Patrick's Day annually exposed the specific realities of a largely English and Scottish production team in Belfast broadcasting to Catholic listeners, while some Unionists questioned why the saint's day was being marked at all.[37] This was, of course, a singularly troubling issue, but it exemplifies the central problem: just as some Ulster listeners might positively relish the music of pipes and jigs, others would find them alienating and offensive.

No amount of well-intentioned 'voices out of the air' could truly paper over the gaps and fissures in the walls which both housed and divided British society. Even so, the new calendar of events was at least providing some sense of a shared national identity available to all, and listeners were certainly hearing more about the cultures of different parts of the country than had ever been the case in the pre-wireless years.

⁂

The year 1932 was eventful for broadcasting. As the king's message had pointed out, the BBC's reach had recently become very much wider, with the Empire Service (later the General Overseas Service, now the World Service) going on-air just a few days earlier, at 9.30 a.m. (and then repeated five times to reach different parts of the globe) on Saturday 19 December. Initially an English-language service, its specific mission was to link up British speakers in the far reaches of the Empire. Tellingly, an experimental broadcast back in 1930 had led with the British weather forecast, as though information about whether it was raining in Ramsgate or sleeting in Scarborough was all that homesick listeners might truly crave. It was only in 1938 that broadcasting began in a variety of other languages and became a hugely important cultural carrier in its own right, as well as a powerful voice of British soft diplomacy. Back then, in his opening broadcast, John Reith emphasized its unifying intent:

> If, in general, as is our hope, the several far-scattered units of the family may be drawn closer together, then our efforts, which culminate in a preliminary sense today, and definitely in the afternoon of Christmas day, when His Majesty the King will speak for the first time to the Empire as a whole, then, as I say, our efforts are amply rewarded.[38]

⁂

The expansion was not just on-air: the BBC had also invested in bricks and mortar (or, rather, Portland stone), and 1932 saw the opening of its first purpose-built home – Broadcasting House – in a central London site between Oxford Circus and Regent's Park. The iconic art deco building, both sleek and substantial, famously cut into Portland Place like a great liner coming into port (or perhaps setting out on an

Broadcasting House opened in 1932 as the BBC's modern new home in the heart of the capital.

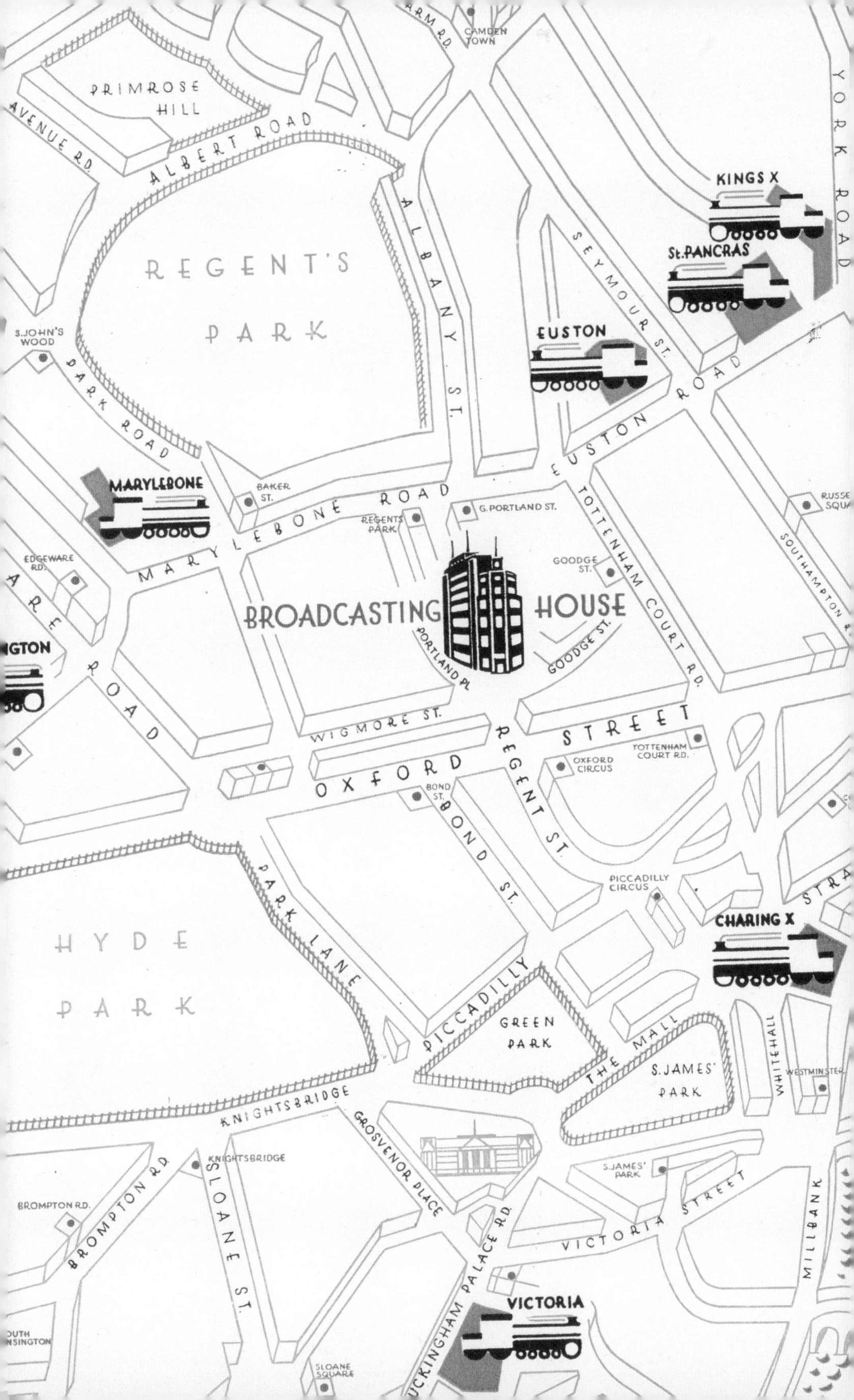
CAMDEN TOWN
PRIMROSE HILL
AVENUE RD.
ALBERT ROAD
YORK ROAD
KINGS X
St.PANCRAS
REGENT'S PARK
ALBANY ST.
SEYMOUR ST.
EUSTON
S.JOHN'S WOOD
PARK ROAD
EUSTON ROAD
MARYLEBONE
BAKER ST.
MARYLEBONE ROAD
G.PORTLAND ST.
REGENTS PARK
TOTTENHAM COURT RD.
SOUTHAMPTON
EDGEWARE RD.
GOODGE ST.
BROADCASTING HOUSE
PORTLAND PL.
GOODGE ST.
WIGMORE ST.
STREET
OXFORD
TOTTENHAM COURT RD.
OXFORD CIRCUS
REGENT ST.
BOND ST.
BOND ST.
PARK LANE
PICCADILLY CIRCUS
CHARING X
HYDE PARK
PICCADILLY
GREEN PARK
THE MALL
S.JAMES' PARK
WHITEHALL
WESTMINSTER
KNIGHTSBRIDGE
GROSVENOR PLACE
KNIGHTSBRIDGE
BROMPTON RD.
SLOANE ST.
S.JAMES' PARK
BROMPTON RD.
BROMPTON RD.
VICTORIA STREET
BUCKINGHAM PALACE RD.
MILLBANK
VICTORIA
SLOANE SQUARE

uncharted ocean). Visitors who were unfamiliar with the area were told to look out for a building like a ship with the wrong kind of windows.[39] The *Architectural Review* of 1932 dubbed it 'the new Tower of London',[40] alluding also to the Norman castle's full title of Her Majesty's Royal Palace and Fortress. Many newspapers commented on its palatial proportions – its mile of corridors, its 840 radiators. Most drew particular attention to its technical specifications and the idea that improving the listening experience was at the heart of this colossal building project. 'Radio Tower of Marvels: Latest in Everything at Broadcasting House'[41] was a typical headline, though one reporter singled out for comment a rather different design feature – the fact that 'the captain's cabin', Lord Reith's personal office, contained the building's only fireplace.[42] As listeners sat by their own single fireplace, they could be assured that a new era of wireless broadcasting was being kindled in this future-facing new home of the BBC.

Meticulous attention to detail had accompanied every aspect of the building project. The music department, for example, had conducted a series of blind tests in order to decide which pianos to purchase for their many studios. When the British manufacturers Challen were eventually chosen, they swiftly put out a series of advertisements asking, 'Why not investigate this triumphant British instrument for your own home?' In response, the more modest Monington & Weston company began to promote its instruments as ideal for 'the restricted homes and purses of today'. Soon the German company Bechstein took up the challenge with a rather convoluted statement that this was 'the piano used by the famous when broadcasting for recording at their recitals in their homes'.[43] My parents were given a 'baby' Bechstein for a wedding present in the late 1940s, and it was on this instrument that I first bashed out 'The Bluebells of Scotland', so the piano wars had not entirely been lost even in the late 1960s.

Broadcasting House was and is a magnificent edifice, but the original internal decor of some of the studios was perhaps a bit too literal and old-fashioned in a rather oversymbolic way. Listeners could now see this for themselves in illustrated articles in popular magazines: the studio from which religious services were broadcast featured three ecclesiastical arches, the central one lit to produce an effect of infinite distance above a sort of altar; one of the talks studios resembled a book-lined eighteenth-century study, with a portrait of George Washington over the mantelpiece and solid Victorian brown furniture.[44] Like the exterior building, wireless was such a fabulously exciting modern thing. In an increasingly competitive world, though, there was just a suggestion here that the BBC might be struggling to keep up with the times, and to respond to what the listeners, with their now well-established understanding of radio, were demanding.

The OFFICIAL RULE BOOK *of* The League of Ovaltineys

WARNING! This book is *strictly private*. It contains confidential rules and secrets that are intended for members of The League of Ovaltineys only. If lost, the finder is requested to return it, WITHOUT OPENING, to the owner whose name is on the back cover.

EIGHT

WE ARE THE OVALTINEYS

> I don't know about the public taste, but I know what I can do with and what I can't. I have a total distaste for the 'now I kiss your hand, madame' style of thing. But there's a Lyon's tea man comes in here – quite a man about town sort of man – spends his holiday at Southsea. Now he did four separate shows all the time he was there, and now he's back, he has his wireless raving out dance tunes all the time. The wireless has certainly intensified a desire for music. Mind you, they're afraid of good stuff, most people ... Sibelius Finlandia, now I believe we'd come home from a race meeting to hear that.
>
> Mr James, grocer, Bristol[1]

As soon as an additional bank holiday was declared for Monday 6 May 1935, preparations got under way across the country. Rural communities began to organize carnivals and garden parties, sports events for the children, suppers for the elderly, pageants and variety concerts, while many towns and cities opted for banners, bunting and brass bands. Since no British monarch had ever celebrated a silver jubilee, there was no precedent for the formal ceremony, but the BBC was clearly going to be in on the act. As the day itself drew to a close, Gerald Cock, the BBC's first Director of Outside Broadcasts, had reason to be proud of his success in sharing with listeners the royal carriage procession to St Paul's and the service of thanksgiving which had been captured live via ten microphones laid along the

Few events shook the self-confidence of the BBC as much as the advent of this winsome group of fictional children who first appeared on Radio Luxembourg in 1935.

WILLS'S CIGARETTES
RICHARD TAUBER
RICHARD TAUBER. Richard Tauber was born in Austria, his father being an actor and conductor of the State Theatre at Chemnitz. He studied music at the Frankfurt Conservatoire, becoming a conductor at the age of eighteen. About this time he discovered that he had a fine voice, and abandoned conducting to make his début in "The Magic Flute." Tauber had very little vocal training and can really claim to be self-taught. He practises his songs with his own accompaniment at the piano, for he is a pianist of considerable ability. His first appearance in operetta was in Franz Lehar's "Frasquita," and he has been the greatest exponent of Lehar ever since. (No. 22.)
PAUL ROBESON. Singer, actor, lawyer and athlete, Paul Robeson (who was the son of a well-known coloured preacher) was born in New Jersey, U.S.A. in 1898. He was educated at Columbia University and studied law. Playing negro rôles in two Eugene O'Neill plays, Robeson created a sensation ; this started him on a new career, and from the stage he went to the concert hall. His first concert, in 1925, was a great success, and that year he came to London to play lead in "The Emperor Jones," and again in 1928 to sing "Ole Man River" in the "Show Boat." Since then he has made his home in London. He first broadcast here in 1929. (No. 23.)
WILLS'S CIGARETTES
PAUL ROBESON
WILLS'S CIGARETTES
DENNIS NOBLE
DENNIS NOBLE. This very fine
singer was educated in the Cathedral
of Bristol, his native city, and became a
there. Later he joined the Westminster
staff, becoming lay vicar of the Abbey and
baritone of the choir. In that capacity he
his first broadcast on the night of the
the Duke of York, when the Abbey
packed into a tiny B.B.C. studio at Marconi
and sang a special anthem. He has since
hundreds of times in all types of programmes,
musical comedy to "Sea Drift" at the
tival. An enthusiastic cricketer, he is a
of several London cricket clubs.

Wills collectible cigarette cards had featured 'Ships and Sailors', 'Aviation', 'Lucky Charms', 'British Butterflies', 'Famous Golfers' and 'Garden Flowers'. 'Wireless Celebrities' were a natural mid-1930s addition.

route, a further seventeen in the cathedral itself and the talents of a well-drilled team of presenters, producers and sound engineers.[2] In the evening, there was a 'Radio-Dramatic Survey' of the king's twenty-five years on the throne, tributes from across the Empire and from the prime minister, Ramsay MacDonald, and a broadcast by the king himself, in which he gave thanks to his 'dear people'. At nine o'clock Rudyard Kipling made a speech; at ten thirty the Poet Laureate, John Masefield, read his Jubilee poem, 'A Prayer for King and Country'. There was also community singing – with listeners encouraged to participate by their own firesides – and a round-up of the local celebrations of the day from all parts of the country 'told in their own words and through their own spokesmen'. What many listeners must most have enjoyed, though, was the prime-time slot between

WILLS'S CIGARETTES
ANONA WINN
NONA WINN. Charming and petite Anona
Winn is one of the busiest and most accom-
shed of radio revue artistes. Born in Sydney,
was reading for the Bar when Melba heard her
ging one day and offered her a scholarship.
er singing in the leading concert halls of Mel-
rne and in musical comedy, she came to London
appeared in "Hit the Deck"; she began
adcasting in 1928. In Australia she was the
artiste to broadcast, and was in the first revue
televised. Anona Winn has composed several
ular songs, including "What more can I ask?"
(No. 25.)

KATE WINTER. Known as "the silvery-voiced soprano of the air," Kate Winter originally had leanings towards the piano, on which instrument she wanted to become a professional musician, but instead she became a school teacher. Her cousin, Grace Day-Winter, began her voice training; later she studied at the Royal College of Music and took lessons from Ivor Foster the famous baritone. On her marriage she gave up school teaching and took up singing. Sir Henry Wood coached her and later she was able to sing under him at a Promenade Concert. She began broadcasting in the very early days at Marconi House, and has been a regular B.B.C. singer ever since. (No. 26.)

MARJERY WYN. Born in Leeds, this golden-haired actress came South to begin her career in a concert party at Westcliff-on-Sea. She returned to Yorkshire as principal girl in the "Babes in the Wood" at Huddersfield. Musical comedy work followed, and in the revival of "The Lady of the Rose" opposite Harry Welchman, she caused a furore by appearing as a West-End leading lady without having previously had West-End experience. When "Mr. Cinders" went on tour she took Binnie Hale's part; she was also in the revival of "The Quaker Girl" and with Stanley Lupino in "Hold My Hand." Marjery Wyn first broadcast three years ago. (No. 27.)

ten past eight and nine o'clock, when a civic celebration was relayed live from the New Corn Exchange in Brighton. The 'All-Star Variety' cast included a troupe called The Dancing Daughters; Norman Long, the hugely popular entertainer who had been broadcasting – with one significant intermission – since the very beginnings of wireless, with a version of his act 'A song, a joke and a pianoforte'; comedy duo sisters Doris and Elsie Waters; Australian-born actress and singer Anona Winn; and music from Jack Payne and his Orchestra. These fifty minutes were a well-judged addition to the formal events of the day and a sure sign that the BBC was now heeding the call, which many listeners had been making since at least the beginning of the decade, for more popular output.

⁂

At the start of the previous year, when the launch edition of *Radio Pictorial* had appeared on newsagents' shelves, it had been joining a wide range of magazines already catering for different wireless interests:

WILLS'S CIGARETTES
NORMAN LONG
NORMAN LONG. Norman Long might still have been an insurance agent had he not discovered during the War that he could amuse the soldiers in his regiment singing funny songs at the piano. He was the first entertainer to broadcast at Marconi House in 1922, and he broadcast again at the opening of the Savoy Hill studios. Another "first" was his broadcast at the Royal Command Performance in 1927—the first Command Performance to be heard over the air. He is one of the eligible bachelors of radio, although his broadcasts have brought him many offers of marriage. He plays golf and was president of the Vaudeville Golfing Society. (No. 40.)

ELSIE AND DORIS WATERS. Elsie and Doris Waters are the originators of the famous "Gert" and "Daisy" characters. They were born in London and with their four brothers formed a family orchestra when young. Elsie studied the violin and was a pupil of Albert Sammons. First broadcasting in March, 1927, they have never once repeated a sketch. They have played before the Princess Royal and have been presented to the Prince of Wales. The sisters write all their own material, both music and lyrics. Elsie is fair and the "Gert" of the partnership, Doris is dark and plays "Daisy." They usually appear before the microphone in their "char-lady" clothes. (No. 41.)
WILLS'S CIGARETTES
ELSIE AND DORIS WATERS

WILLS'S CIGARETTES
JANET JOYE
JANET JOYE. "The girl with a
sonalities," Janet Joye dreamed of
Shakespeare, and studied at the Royal
of Dramatic Art. She might have been
Shakespearean actress—if Shakespeare
but the Bard wrote no leading lady rôles
as tiny as Miss Joye. So she became a
mimic. She learns how to imitate cats
from her Manx cat, Mr. Tom Fuss, and
Irish terrier. Janet Joye writes all her
erial and prefers creating characters to
living ones. Her recent series of
were a popular feature of the
(No. 42.)

magazines for scientifically minded readers, manuals for those seeking practical guidance on building their own sets, and of course the *Radio Times*. The front cover of *Radio Pictorial* announced that this was to be a 'New Weekly Pictorial Magazine for Every Radio Listener'. At 2*d*, it cost the same as the *Radio Times*, but a glance through its pages quickly indicated the target readership, with an emphasis on dance music, showgirls and gossip about the home lives of the stars.[3] Page 5 features an imposing photograph of Sir John Reith, accompanied by a message wishing the new venture well, and BBC stalwarts such as Head of Drama Val Gielgud and the popular gramophone critic Christopher Stone also contributed articles, although endorsements like this didn't last long. There was nothing remotely radical about *Radio Pictorial*, but, as the next issues were gradually to reveal, this was a landmark publication, appealing to the growing number of dissatisfied listeners who were either wondering about retuning their wireless sets beyond the BBC or were already doing so. The title of

WILLS'S CIGARETTES
BILLY MAYERL
LLY MAYERL. "The pianist with lightning in his fingers" has been playing the o all his life, and at the age of seven he got into race for "jazzing" Beethoven and Grieg while ying at the Trinity College of Music. At the of twelve he played the Grieg Concerto at en's Hall and was hailed as an infant prodigy, his career was interrupted through family misnes. While playing the piano at a local cinema Mayerl discovered that there was more money ncopation than in symphonies. He joined the y Orpheans and became famous in a night; as been broadcasting ever since. (No. 43.)

CARROLL GIBBONS. Born in Massachusetts, U.S.A. Carroll Gibbons studied harmony and piano technique at the New England Conservatory of Music in Boston. In this country he was the solo pianist of the original Savoy Orpheans, and now directs the dance orchestra at the Savoy. A bachelor and only just over thirty, he plays golf and admits he plays it badly, but likes driving fast cars. His now famous tune "On the Air" is the signature tune of Rudy Vallee, the famous American dance band conductor, who is a great friend of Carroll Gibbons. (No. 44.)
WILLS'S CIGARETTES
CARROLL GIBBONS

WILLS'S CIGARETTES
CHARLIE KUNZ
CHARLIE KUNZ. The Casani Club Orchestra directed by Charles Kunz is one of the most popular dance bands on the air, and Charlie himself as a pianist is even more popular. His signature tune, as most listeners know, is "Clap hands, here comes Charlie!" which is based on the song "Here comes Charlie"—popular among dance bands when he came to England twelve years ago. Born in Pennsylvania, Charlie, after trying many ways of making a living, found one to suit him—a night job playing the piano at a local hotel, which he combined with delivering early morning milk, going straight from the hotel to the milk round in his evening clothes! (No. 45.)

the magazine's cartoon strip was 'The Twiddleknobs', and this name for the comic family alone reveals what was now a fast-growing trend.

The exploratory twiddling of dials and tuning of receiving gadgets had been the activity beloved by amateurs back in the days of Morse code – amazingly, not much more than a decade earlier. Even in the early days of broadcasting, wireless owners recalled patient searching for a signal. It was principally in order to leave behind this concept of random eavesdropping that John Reith had specifically rejected the term 'listening in' and replaced it with the more purposeful 'listening'. For him, all listening should be purposeful and the provider of all listening was the BBC. In his singularity of purpose – the very defect of his finest quality – he imagined that others would feel the same. They did not.

For a start, the BBC was only available at some hours of the day and not at all at night. As Mr James from Bristol recognized, it is often at night that the gift of radio can be greatest:

> My father died 10 years ago. He had anaemia and he slept very badly. About 3 in the morning, he'd say 'Anything on the wireless?' And I'd fiddle round till I got him some dance music from somewhere abroad. I'd make him a bit of toast and the time'd pass for him. Now where would he have been without the wireless?[4]

This poignant 1939 recollection of a 1929 experience perfectly captures how the comfort of such nocturnal companionship was first discovered, but listeners were not only seeking out alternatives at night.

The biggest problem was on Sunday, which was for many people the only whole day of the week when they were not at work. Reith was an utterly intractable Sabbatarian, stating in his 1924 autobiography that 'the secularising of the day is one of the most significant and unfortunate trends of modern life'[5] and imposing this view upon BBC output. The fact that, in 1926, he questioned whether the traditional English folk song 'Oh No, John' was suitable for Sunday broadcast gives an idea both of his rigidity and of the sort of limited output listeners heard on-air.[6] Take a look at the Sunday listing in the first issue of the *Radio Times*. The morning was silence, on the assumption that listeners would be attending their own place of worship. From three o'clock in the afternoon there were organ recitals, followed by the RAF band playing mostly sacred music, the *News* and then close-down. A decade later, on the Sunday of the week that *Radio Pictorial* first appeared, the schedule was far fuller, but had altered little in tone. Apart from a 10.30 weather forecast, there was still silence until 12.30 when the Scottish Studio Orchestra performed a concert. A range of classical recitals followed, as well as talks on 'British Art', 'Pillars of the English Church', episode 48 (!) of a series entitled *Readings from Classical Literature*, and a programme for children titled *Joan and Betty's Bible Stories*. Then there was a lengthy religious service and *The Week's Good Cause*. The most secular number in the early evening concert relayed from the Park Lane Hotel was rather primly entitled 'Invitation to the Dance', then an *Epilogue* and close-down at 10.30 p.m. Some items were appealing in their own right, but it was hardly an appetising menu for a whole day of listening. In an era when the influence of the church was waning and attendance

Radio Pictorial was a weekly magazine independent of the BBC, providing listeners with a far wider range of choices when it appeared in January 1934.

at services dropping – there is documentary evidence of this in Barton Hill – many listeners were finding such Sunday offerings heavy going. They might stay loyal during the week – most did – but on Sundays a remarkable two-thirds of listeners, many of them young working-class women, were retuning their wireless sets. One individual account provides insight into this mass defection.

Just at this time, Benjamin Seebohm Rowntree, the second son of the Quaker cocoa and chocolate manufacturer Joseph Rowntree, was conducting the middle of his three sociological studies into poverty in York. Although only a very few samples survive of his *Wireless Listening Habits Survey*, these show a consistent preference among listeners for light music, dance bands and variety. It's a response to the questionnaires about weekly leisure-time activities which provides the most vivid glimpse into Sunday life:

> No one in the house gets up before 10 a.m. The wife immediately starts getting the dinner ready and we both have a cup of tea. After dinner we both laze about, either reading or sleeping. Sunday is the worst day in the week, absolutely dead. After tea, we roll up the carpet, find a foreign station on the wireless giving a dance band, and we dance most of the evening, sometimes playing cards for an hour before going to bed.[7]

Nobody up until 10 a.m., no religious observance, 'lazing', 'sleeping', 'dancing', 'playing cards' – apart from the reading, it's hard to imagine a Sunday schedule less likely to gain the approval of John Reith. And now, to make matters far worse, *Radio Pictorial* was providing listeners with a guide which expressly publicized the continental commercial stations. On that first Sunday, there was a constant stream of light music, dance bands and operetta broadcast from Breslau, Berlin and Brussels, Toulouse, Vienna and Warsaw – all available at the twist of the dial. A growing number of listeners had now invested in the new, more powerful superheterodyne sets, which made finding these foreign stations all the easier.

The increasing popularity of one station above all others was making Reith fume: even he could not ignore what he termed 'the monstrous stuff from Luxembourg'. A casual comment from one of the York families that they had listened to the station for 'six hours (continuously)'[8] would particularly have infuriated him, as he had always voiced a deep distaste for what he called 'tap' listening on any day of the week, let alone Sundays. Meanwhile, Winifred Gill scribbled a note on the corner of a page in her reporter's notebook acknowledging that schoolchildren had mentioned listening to Radio Luxembourg – though mysteriously this never made it into the published pamphlet.[9]

Back in 1922 the founding principles of the BBC had been laid down in opposition to the American model with its regular advertisements. Like other European commercial stations, Luxembourg too was dependent on advertising. As its reach grew, many popular brands bought airtime, and their names rang out repeatedly. Programmes were short and sometimes the *Radio Pictorial* billing simply gave the sponsoring brand as the title: 'Carter's Little Liver Pills', 'Maclean Stomach Powder', 'Horlicks Tea Hour', 'Rinso', 'Outdoor Beauty Girl', 'Harry Peck & Co. Ltd Makers of "Chix" Chicken Broth Cubes'... A number of the on-air voices were already household names as there had been some serious poaching from the BBC. Christopher Stone, for example, now wrote a weekly column recommending other English-language programmes from the Continent and he also hosted his own show. Before long, even though his photo byline clearly showed him smoking a pipe, his image appeared in the pages of *Radio Pictorial* in an advertisement that was also a recommendation for listeners to tune in on Sunday and hear him presenting the 'Will's Star Cigarette' programme. This kind of muddying of the waters between editorial content, product placement and outright advertising was on the increase. Later in the decade, the 'Halifax Toffee King', John Mackintosh, sponsored a programme entitled *That Reminds Me*, which was designed to promote a brand-new line of novelty chocolates and toffees. A few precious minutes survive of the audio in which the presenter, known as 'Melody Mac', is in hesitation-free, clipped-vowel full flow:

> Now if you would like to join in this programme, just send us the name of the melody that brings back certain memories, and then tell me the memories in about a hundred words ... if your story is broadcast, you will receive a prize of half a guinea. You'll love Quality Street, such variety, such novel succulent flavours intermingled with the good old favourites we never want to miss. There's really no better entertainment in the world than listening to the radio with a tin of Macintosh's Quality Street at your side. That reminds me, listen again next week at the same time to another of these interesting programmes and don't forget to send in your story and the tune that reminds you of it.[10]

Even the Rowntree company was soon advertising on Luxembourg. The views of Seebohm Rowntree on 'unimproving recreation' were on a par with those of John Reith, but he had to juggle his Sabbatarianism with his day job as one of the most powerful businessmen in the country. Letters survive between members of his circle expressing dismay at hearing the family name in advertisements broadcast on a Sunday.

Cigarettes, toffees and chocolates were small weekly luxuries consumed by many adult listeners, but in January 1935 Luxembourg launched a programme sponsored by a brand of malted drink which was specifically targeted at the all-important next generation of listeners. Its syncopated theme tune was so sensationally catchy that it was wheeled out again for a shamelessly nostalgic 1978 television advertisement, with the results that, to this day, a surprising number of people find they can still sing along to it:

> We are the Ovaltineys, little girls and boys,
> Make your requests, we'll not refuse you
> We are here just to amuse you
> Would you like a song or story?
> Will you share our joys?
> At games and sports we're more than keen
> No merrier children could be seen
> Because we all drink Ovaltine
> We're happy girls and boys.

The first outing of *The Children's Special Half-Hour Entertainment Broadcast especially for the League of Ovaltineys* was trailed by several weeks of print advertisements, which are now mostly remarkable for their lack of subtlety: one shows a photograph of a small boy holding a mug in his chubby hands and declaring 'Oh Mummy – it's lovely!' before informing readers that 'Ovaltine is without equal as the daily beverage for building up robust health and abundant vitality'. Below this half-page panel sits a framed announcement reminding readers not to forget to listen to the forthcoming *Special Children's Programme*, '5.30–6.00 p.m. on Sunday next from Radio Luxembourg', with the promise of music from (BBC favourites) Jack Payne and his band, as well as Christopher Stone. Then, on 8 February 1935, came the half-hour weekly *Concert Party*. Harry Helmsley, the compere and composer of the theme tune, already had a name from his music-hall days for his children's impersonations. These had now coalesced into a family of four children – purportedly his own – named Johnny, Elsie, Winnie and baby Horace, whose incomprehensible babble frequently needed interpreting by the cleverest of the children, Winnie. 'What did Horace say, Winnie?' soon became a popular catchphrase, bringing a sense of complicit delight to many listeners.

For listeners still loyal to the BBC on that Sunday, the main children's offering was a church service, as *Joan and Betty's Bibles Stories* had, by now, thankfully been shelved. The broadcast which coincided precisely with the new Ovaltiney half-hour happened to be a special adaptation of *Troilus and Cressida*, which even the *Radio Times* acknowledged was 'Shakespeare's most mysterious play'. With Troilus's opening words – 'Call here my varlet, I'll unarm again...' – the crackle of static must have been audible from street to street as listeners twisted their dials to find the heavily trailed new programme which unambiguously promised 'just to amuse you'.

The *League of Ovaltineys* was packed with product placement, brief songs, snatches of stories and canned laughter, all performed by children from the Italia Conti Theatre School, where elocution classes seem to have filled many hours of the curriculum. Listening now, the forced merriment sounds pretty grating, but it's not hard to understand why listeners – adults as well as children – might have switched over to its uncomplicated, light-hearted spirit. This was not necessarily because they had no interest in Shakespeare but because

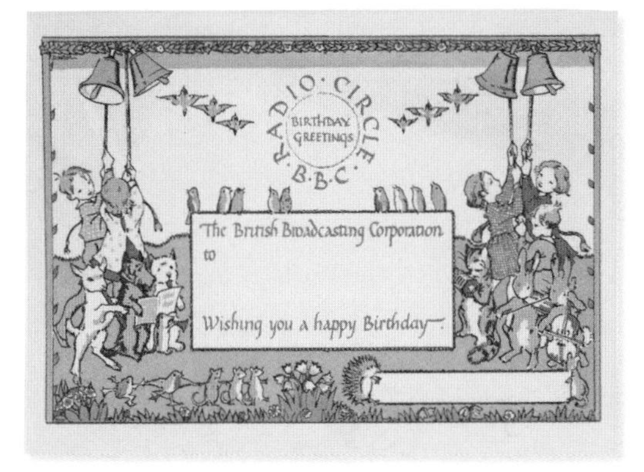

The endearing design, by Dorothy Hutton, of these membership cards for young BBC listeners perhaps began to seem a little staid as alternative programmes became available on the Continental commercial stations.

they might not have felt that Elizabethan drama was quite what they fancied as the weekend drew to a close and many began to think of the cold, anxious week ahead.

The popularity of the new strand was also sealed in other ways. The *Ovaltineys Concert Party* was a regular Sunday fixture; membership of a club rewarded listeners with a badge, rule book and, later, a comic; there were frequent competitions, though participation invariably involved a further purchase. Of course, the BBC's own *Children's Hour* had been immensely popular from the start – with adults as well as children – but Sunday was the one day of the week when it was not broadcast. It too boasted a club and badge. The Radio Circle's cream-coloured certificates, featuring children, fairies and animals drawn by the painter and calligrapher Dorothy Hutton, were simple and dainty but perhaps beginning to look a touch old-fashioned. Initially, membership had promised an on-air birthday call-out ('Many happy returns to Jonny Green of Harbourn, and if he will look in the coal scuttle, he will find a present'[11]), though by 1933 this had had to be discontinued because of the sheer number of requests – evidence of its success but also, sadly, of a gradual distancing from its target audience. Certainly, by this time some of the heady laughing gas of the earliest days had been sucked out of programmes. It was said, for example, that one of the Aunties had been reprimanded by John Reith for adding an impromptu 'Yo ho ho and a bottle of rum' to an episode about pirates. In 1934 the editor of the *Radio Times* defended the various changes by arguing that 'the spontaneous fooling of the Aunts and Uncles has given place to programmes that even the precocious child can enjoy. The Children's Hour has shown that broadcasters can enjoy broadcasting without being undignified.'[12] Even Hilda Matheson had argued that the strand did not need to 'depend upon condescending heartiness nor silly puerility to be popular'.[13] Both seemed to overlook the fact that many children adore 'silly puerility', and particularly enjoy moments when adults are at their least dignified. With her satirical eye, E.M. Delafield summed up *Children's Hour* in not entirely complimentary terms:

> From the suave and carefully condescending voices of kind ladies and gentlemen, little children, if so minded, may hear about distant countries, about the private lives of film stars, the habits of African elephants, the importance of cleaning their teeth, and the difference between a minim and a quaver.[14]

Unforgettable – but perhaps too unkind. The BBC was offering children their own programme on six days of the week, and without the constant references to commercial sponsors. Public-service broadcasting could almost be defined as providing a balanced diet as opposed to a milky bedtime drink advertised with the frankly disingenuous promise that it contained 'no household sugar', but, during these middle years of the 1930s, the corporation was increasingly struggling with its menu.

❧

The commercial imperative now entailed extensive audience research by the continental stations, but the BBC resisted this route until 1936. A 1930 classic Heath Robinson cartoon provides a brilliant triple image suggesting that such data might be gathered by less conventional means: 'A Meteorologometer for Testing the Accuracy of Weather Reports'; 'The Epiloguemeter for Testing the Punctuality of the Epilogue' (the final words before close-down) and 'A Simple Device for Testing the Attractiveness of the Children's Hour'.[15] This final mechanism depicts three tots being lured from a large amplifying horn by the smell of freshly baked buns. The food analogy is useful in trying to understand more fully why listeners were turning away from the BBC.

When George Orwell set off, in 1936, to experience first-hand what social conditions were like among unemployed people in Wigan, Barnsley and Sheffield, perhaps his greatest achievement was his imaginative leap in trying to understand how the reality of their everyday lives might affect certain choices. A highlight of the book is his analysis of an article in the *New Statesman* about the minimum weekly sum on which an individual could stay alive, and the series of economically precise but emotionally heartless readers' letters that followed. Of course, he agrees, it would be possible to survive on raw carrots (nothing spent on fuel for cooking) and wholemeal bread (far healthier than the soft white variety), but

> when you are unemployed, which is to say when you are underfed, harassed, bored and miserable, you don't want to eat dull wholesome food. You want something a little bit 'tasty'. There is always some cheaply pleasant thing to tempt you. 'Let's have three penn'orth of chips! Run out and buy us a twopenny ice-cream! Put the kettle on and we'll all have a nice cup of tea!'[16]

Radio critics also used food analogies when writing about the BBC. One fairly generous article referred to the broadcaster as providing 'a universal dinner',[17] another to the idea that Reith was slipping 'the

Heath Robinson's cartoons depict various tests for the effectiveness of radio broadcasts. They were published in the *BBC Yearbook 1930*, six years before the establishment of a rather more conventional actual Audience Research Unit.

Gregory powder of uplift into the jam of recreation'.[18] The name of this patent medicine for indigestion, acidity, heartburn and irregular bowel movements was universally familiar at the time, making the metaphor neither appetising nor complimentary.

Criticism of the BBC offering was nothing new. The radio manufacturers frequently called for lighter, brighter output – more variety, more popular music, more sport – while wireless critics with 'man of the people' bylines such as 'Jack Broadcaster' or 'Wavelength' mounted regular attacks, including a call for the BBC to be rebranded as the 'British Boredcasting Corporation'.[19] Of course, both the manufacturers and the popular press had their own agendas. Even so, there's no doubt that, in Orwell's terms, many listeners were now finding the BBC's offering far less tasty than that of its rivals, far too heavy on the wholemeal bread. It was a complaint with a long history.

The idea that the function of broadcasting might be 'a means of education and entertainment' had been enshrined in the royal charter of 1927. The renewal of this charter ten years later was to carry with it the iconic promise of 'information, education and entertainment'. As a mission statement this was unimpeachable. The question, during these middle years of the 1930s, was increasingly one of priority, proportion and balance in its on-air schedule.

The educational role of broadcasting and what informally became known as 'cultural uplift' had always been Reith's priorities. His policy 'to bring the best of everything into the greatest number of homes' was admirable in principle, but his lack of self-doubt was not unproblematic:

> It is occasionally indicated to us that we are apparently setting out to give the public what we think they need – and not what they want – but few know what they want and very few what they need... In any case it is better to overestimate the mentality of the public than to under-estimate it.[20]

In fact, parts of this famous statement later turned out to contain some seeds of truth, though not on his paternalistic terms, and not yet.

For the present, even listeners who met Reith's high criteria might encounter practical problems, since most households contained only one wireless set. As a writer in the *Huddersfield Daily Examiner* pointed out in 1930:

> The more earnest member of the family who is anxious to listen to some educational talk is likely to meet with the opposition of those of a less serious turn of mind, who, by their very insistence and numbers, will probably in the end prevail and switch over to the nearest station offering the alternative allurement of jazz.[21]

The BBC had in fact already recognized this issue. For some time it had been developing a pioneering model of wireless-based education for 'the plain man' which was expressly designed to take place outside the home. At its heart were 'wireless discussion groups', which would meet in the early evenings, once a week, in village halls, libraries, schools, or clubs for working men or the unemployed. Their purpose was shared listening, among like-minded people, to series of six or twelve half-hour talks on a given topic, followed by a discussion under the guidance of a leader chosen from among them. Subjects were wide-ranging and timely: 'The New Spirit in Literature', 'Science and Civilization', 'Can Democracy Survive?' 'Music Old and New', 'Must Britain Starve?' 'The National Character', 'From Tolpuddle to TUC', 'Among the Roman Roads', 'Child, Parent and Teacher', 'Design in Modern Life'...

For this major plank of the BBC mission there was a director of adult education, a general council, a network of regional councils, conferences and summer schools. Many of the staff involved shared Reith's belief in the power of education, although there was something kindlier in how they actually attempted to roll it out. The advice published in two contemporary booklets is encouraging, accessible and, above all, realistic in tone: it was important that participants were comfortable and that reception was satisfactory; it was sensible to meet a little before the broadcast so that some informal chat could take place; it was vital that the group leader had enough authority to rein in the most self-confident while encouraging younger or shyer members to participate. These booklets featured photographs

of groups (mostly men) in action in a barn, a chapel in Liverpool, a remote Scottish croft and a somewhat utopian cottage porch outside which two elderly pipe-smoking men in suspiciously clean suits and gaiters sit, hats in hands, headphones on ears, listening attentively (an image used more than once in BBC publications). A leader of a group held at a Men's Evening Institute in the East End of London wrote a touching letter to thank the BBC for giving him access to first-class scholarship and opinion, of 'throwing open the windows … and giving us draughts of fresh air'.[22] He was not alone in his enthusiasm, which

From the late 1920s and throughout the 1930s, listeners could buy versions of educational radio talks. These inexpensive pamphlets (*left, above & overleaf*) were often handsomely produced and featured commissions from leading artists such as Eric Ravilious (*left*).

must have been gratifying for the teams working on such output. As early as May 1927, the *Radio Times* carried a striking photo on its front cover showing two young men listening to educational output. The caption read: 'Students of the New Era: Today we can no longer speak of the loneliness and isolation of country life. Radio has changed all that. The youth of the countryside can now receive through wireless the best that the university has to offer.'[23] But Reith's hope of a social revolution triggered by such groups soon faded: the steep line on the graph showing enthusiastic initial uptake soon plateaued, with some participants complaining that the content was too dry, the delivery too fast, the tone too superior. These were accusations which had long been levelled at general talks.[24] One 1933 survey, admittedly conducted by a group of manufacturers, suggested that 80 per cent of the public who had bought their sets on hire purchase ignored talks and readings altogether.[25]

The broadcasting of classical music was open to similar criticism: some relished the new opportunity to extend their knowledge; others

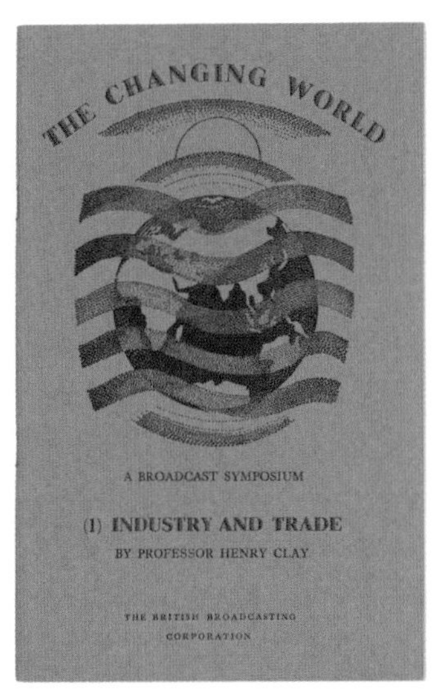

deplored what they perceived as highbrow and improving output. The BBC had in fact struck gold with Sir Walford Davies, whose talks on *Music and the Ordinary Listener* and *Everyman's Music* had a lightness of touch which soon turned him into a household name. As one listener later recalled, 'He always seemed to come right into the room with us.'[26] There were Barton Hill residents who observed that the general standard of musical knowledge and appreciation had been expanded by wireless listening:

> Look how you hear people walking along in town humming. You'd have been thought crazy to do that – people would have turned to look at you. A man'll hum to himself in all the roar of a factory. No one can't hear him, he can't hear himself, but he'll do it all the same.[27]

Others shared opinions about a growing interest in what they called 'better' or 'better-class' music. Mr James voiced this as a marked preference for Sibelius's *Finlandia* over dance music, while Mrs Pugsley was clearly proud of her venture into new territory:

> I went to my first concert because of what I had heard on the wireless. Berlin Philharmonic. I noticed it was very full, and several newsboys came straight from work, with bags of papers on their shoulders – the cheapest seats, 3/6. A friend in the building trade notices what the men

> sing or whistle at work. 'One very ignorant sort of chap' has a large repertoire of Wagner.[28]

Mr Hardcastle, though, was a little more measured in his response, with Gill herself underlining the key words: 'I don't think my taste has improved, but I get more music than I ever did, and I am sure I enjoy it more.'[29] The emphasis on 'more' is not insignificant; nor is the confident naming of a range of classical composers who might well have been unfamiliar before the coming of radio. For many listeners, simply gaining access to talks and music from which they themselves could choose was in itself enriching.

Paradoxically, the immense educational impact of the BBC was already undisputed, but at a casual and informal level. Unexpected glimpses of what truly excited and inspired listeners can sometimes be drawn from borrowing figures from public libraries, which were expanding in use at this time. One librarian in Chiswick and Brentford reported that it had become impossible to keep up with demand for Dumas' novel *The Count of Monte Cristo* when it was dramatized on the BBC,[30] while another, from a rural area where standards even of functional literacy were poor, shared her observation that many people simply enjoyed hearing poetry read aloud on the wireless.[31] Listeners' knowledge, understanding and enjoyment of the world was expanding at the same time as many were ignoring or rejecting the BBC's overtly educational offering.

Listeners were also retuning their wirelesses as a reaction against the BBC's unrelenting high moral tone. In the early days the broadcaster had established its brand as something positively modern, a new twentieth-century presence in the parlour, but by the end of the 1920s the puritanical note it struck was sounding positively Victorian (Oh no, John! No, John! No, John! No…). An embarrassing public falling out with one of its earliest on-air talents provides a perfect case in point. In 1919 the entertainer Norman Long had made his first stage appearance at the Lewisham Hippodrome in *A Song, a Smile and a Piano*. In its voracious early months the BBC had picked this up along with many other similar acts, and repurposed it for broadcast, simply giving it a more aurally appropriate name – *A Song, a Joke, and a Piano*. A decade later, in 1932, Long wrote and recorded a new number which highlighted the growing gap between the BBC's sense of decency and the culture of the music-hall tradition from which it had drawn.[32] He began with a few lines of harmless autobiography before approaching the issue of censorship head on:

> Now the BBC
> Once wrote to me
> And said, Dear Norman Long
> We thought you'd like
> To face the mike
> With your piano, smile and song.
> So will you bring your repertoire
> Along for us to see
> To go over it with a pencil
> Of the blue variety?

He continued with a great run of verses, in each of which his choice of song is rejected. Reasons included content considered even mildly offensive:

> So I looked my list of songs up,
> To please the listening throng,
> I said, 'Now here's a good one,
> The Volga Boatman's Song'.
> But they hated vulgar boatmen,
> And discouraged vulgar chat,
> And songs about rude sailors are banned,
> 'We can't let you broadcast that!'

As well as promiscuity:

> So I looked my list of songs up,
> And found an army one,
> There's Something About a Soldier,
> But they said 'That can't be done!
> There's Something About a Soldier
> Would get over nice and pat
> With a sergeant's wife and fourteen kids,
> We can't let you broadcast that!'

And the singularly touchy subject of advertising:

> So I looked my list of songs up,
> The position was getting grave,
> I said, 'Now here's a nautical song,
> A Life on the Ocean Wave'.
> They said 'Life on the Ocean Wave,
> With all those cruises on the mat?
> Why you're advertising the steamship lines,
> We can't let you broadcast that!'

Like all good caricatures, the song bore some relationship to reality, cleverly skewering an increasing list of official dos and don'ts, rules and regulations published for the benefit of would-be entertainers. The fact that the BBC went on to score a spectacular own goal by banning it of course made matters far worse. Now each and every time listeners heard the song performed, they felt actively encouraged to join in a shared joke against the broadcaster. Welcomed back onto the airwaves by the time of the king's Jubilee, was Norman Long making a humorous declaration of his decency by using the rather over-formal term 'pianoforte' in the title of his Brighton act?

As it happened, the song's popularity coincided exactly with Radio Luxembourg's English-language test transmissions to Britain and Ireland. The BBC could ban it on-air but they had no control over dissatisfied listeners defecting to networks which were broadcasting material more in keeping with their taste and sense of humour. This marked a significant moment and true reversal of power: the voices of the listeners were demanding to be heard.

There was no relenting on Sunday listening: only in 1938 (when Reith was about to stand down) did mid-morning broadcasting begin, but more light music and popular drama were gradually appearing on-air. Most importantly, there was one major structural shift: a new department dedicated exclusively to variety, with an energetic leader to drive change.

Since 1927 Eric Maschwitz had been editor of the *Radio Times*, and his first decision on taking up the new role showed how well he understood the sometimes stifling corporate culture. Even as production teams were installing themselves in the recently opened Broadcasting House, he insisted on moving part of his department to a large theatre nearby, where a more relaxed atmosphere could prevail. Very soon, listeners were happily picking this up on-air.

Maschwitz was extraordinarily versatile in his talents. He later gained fame as the lyricist of 'A Nightingale Sang in Berkeley Square' and 'These Foolish Things', as well as co-adaptor of the Hollywood movie *Goodbye Mr Chips*, but he wasted no time in launching his great wireless legacy. From 18 November 1933, *In Town Tonight* provided what the *Radio Times* billed as a half-hour 'topical supplement' in which show-business personalities who were passing through the capital were soon rubbing shoulders with various London 'characters', such as a female chimney sweep, a ratcatcher employed by Madame Tussauds or a constructor of practical jokes for Christmas crackers. The very first episode featured Bette Davis and a crowd-pleasing band made up of some of the biggest names from popular music: Ambrose on violin, Geraldo on the drums, Henry Hall and Carroll Gibbons on piano.

Such casting had a very broad appeal, but the scheduling was clever too because Saturday evenings on Luxembourg were dedicated to French-language broadcasts. The strand was soon considered to be what is now termed 'appointment listening' – a new concept. Early BBC scheduling had been extremely random, disregarding any need to encourage listeners to make a date with certain programmes or to stay tuned. Long-running strands with episodes broadcast at the same time each week now became a priority, and the idea also grew that if listeners had enjoyed one programme they might stay tuned for a later one. This precious 'inheritance effect' is the holy grail of twenty-first-century broadcasters while also providing a sense of belonging and ownership to their listeners. Tellingly, Radio Four announcers today broadcast from what are called 'Continuity Suites' – the name underwriting a promise that the daily and nightly broadcasts will provide a reassuring network for its listeners.

In 1932 the BBC Dance Orchestra, led by Henry Hall, had a hit with a number titled 'Five-Fifteen', the tune based around the notes B–B–C and the title itself a mnemonic for its daily broadcasts. A few years later, another very popular strand was simply titled *Monday Night at Seven*.[33] One proud Bristol mother reported how assiduously her daughter applied herself to her homework, even when the radio was playing, but commented that she would put down her pen and come and listen to this weekly favourite.[34] The biggest draw of all, though, was *In Town Tonight*, which was also the first series to be given a fixed slot. By the end of the decade, an astonishing 20 million listeners were tuning in at 7.30 every Saturday evening.

The entertainer Bruce Forsyth, who was born in 1928, recalled the weekly ritual of his mother, father, brother and sister gathering around the wireless set:

> We'd all rush to get our chairs in position because we couldn't miss the opening, because the opening was the whole mood of the show, and they'd stop the traffic of London and everybody was still, the whole country was waiting to hear what was going to happen on the show.[35]

As well as becoming radio favourites, Tommy Handley and Ronald Frankau appeared in a 1934 film as Mr Murgatroyd and Mr Winterbottom. The choices of hats and ties suggest the class differences of their characters.

Like one of the catchy advertising jingles on Radio Luxembourg, the iconic opening montage to which he refers gave it instant brand recognition: the upbeat rhythm of Eric Coates's 'Knightsbridge March' established at full volume and then held under the sound effects of a train which cross-faded with a ship's horn and plane, the call of a flower girl, the hubbub of the crowd, a newspaper boy announcing 'In Town Tonight' – all suddenly brought to a screeching halt by a London policeman shouting 'Stop!' What followed was wittier still: the familiar tone of a BBC announcer solemnly introducing the programme with words which soon entered the language: 'Once again we stop the mighty roar of London's traffic and from the great crowds we bring you some of the interesting people who have come from land, sea and air to be In Town Tonight!' He closed the programme with a brief command, barked out in an officer-class accent, 'Carry on London!' As a very small boy, Bruce Forsyth imagined that the streets of London genuinely came to a complete halt for half an hour, and that pedestrians around Piccadilly Circus were all stopping to listen. Adults were no doubt less gullible, with their new understanding that this playful piece of audio was a joke between the broadcasters and themselves. Such complicity was the forerunner of much of the radiogenic comedy in the years to come.

Another early Maschwitz triumph was his decision to give a regular wireless slot to the comedy duo *Mr Murgatroyd and Mr Winterbottom*. Billed as 'Two Minds with not a Single Thought', they were in fact a brilliant quick-fire double act whose own very different backgrounds represented the wide spectrum of BBC listeners: Ronald Frankau was an Old Etonian from a well-established Jewish family, and his patrician tone as Mr Murgatroyd was in marked contrast with fast-talking Liverpudlian Tommy Handley's Mr Winterbottom. Among their fifty or more broadcasts, one sketch stands out because of its extraordinarily timely riff on the importance of the listeners (quoted as the epigraph to this book) – whoever they were, whatever their background, level of education, musical taste, religious inclination, or indeed status in myth or reality: 'Yes, Mr Winterbottom, we must always study the listener – the Mr and Mrs Everymans.'[36]

Radio Luxembourg's figures continued to grow and its output diversified so there was some fruitful if hostile cross-fertilization between the rivals. As the BBC at last demonstrated that it was itself capable of listening, the Silver Jubilee reminded listeners how many roles BBC radio had come to play in their lives: not only keeper of the national calendar and guardian of the common ground between populism and public service,[37] but also their most reliable source of information on stories of shared national interest. The king's death less than a year later set in motion an unforeseen sequence of royal events. Hoping to understand what on earth was going on, listeners returned to their homes and switched on their wireless sets.

EKCO

6 Stage Superhet for UNIVERSAL MAINS

Full Automatic Volume Control

MODEL AD 65 10½ GNS

FULL DETAILS OVERLEAF

NINE

THERE'S A WOMAN IN HULL NOT SINGING

> We couldn't quite hear the words of the ceremony and some of the phrases sounded like 'Gawd blimey' and 'swelp me bob'. We all joined in mock interpretations and there was much laughter.
>
> Mass Observation correspondent from South London, 12 May 1937[1]

By ten in the evening on 11 December 1936, the streets of Barton Hill were deserted. Not only in Bristol but across the country, listeners gathered around their wireless sets. After a brief introduction from Sir John Reith, Edward, now simply Duke of Windsor, began to speak in a slow, clear voice reverberant with emotion. He started by swearing allegiance to his brother, King George VI, and went on to explain the reason for his decision to abdicate: 'you must believe me when I tell you that I have found it impossible to carry the heavy burden of responsibility and to discharge my duties as king, as I would wish to do, without the help and support of the woman I love...'

Edward had been on the throne for just 325 days in this turbulent year of three kings. His father had died at Sandringham on 20 January. From 9.30 on that evening, all BBC networks, both at home and across the world, had broadcast silence, punctuated by updates every fifteen minutes. There had been a short service of recollection and prayer for the king at 10.30 and the announcement of his death just

After it first appeared in 1934, the round EKCO radio became a design classic, transforming the wireless set from a humble household object into a modernist icon.

after midnight. Some listeners must have recalled the way in which the death of the new king's grandmother, Queen Victoria, had been officially announced – by telegram, a sequence of public announcements and then via the newspapers – just thirty-five years previously. Now there was the novel experience of sitting by one's own fireside and sharing in the royal deathbed scene as it unfolded in real time. Reflecting a few days later on the scientific achievement of wireless and its ability to act as 'a gigantic public address system across the world', *Nature* magazine reflected: 'The imagination of a poet like the late Mr. Rudyard Kipling might well have been stirred by this theme of waves of emotion encompassing the earth to trace the changes which history has seen in methods of proclaiming to the nation the loss of its beloved King.'[2]

Back in October 1922 Edward had been the first ever member of the royal family to speak on-air, gamely donning a pair of shorts to address a boy scouts' rally at Alexandra Palace in North London.

This set and headphones were old style by 1936 but the experience of listening on the wireless to updates during the dying days of George V was a revolutionary change for a listener who had been born in the middle of the nineteenth century.

During the tumultuous days surrounding the abdication of Edward VIII, there was a return to communal listening outside radio shops and in other public venues, with crowds anxious to hear the very latest news.

At that time, just before the founding of the British Broadcasting Company, when Marconi's invention was still associated with the military and, by association, with the scout movement, 50,000 had heard the speech in person, while troops across the country had gathered around makeshift wireless sets to listen in. Edward had been presented with a brief account of wireless telegraphy and broadcasting, a souvenir booklet, and – at a time when even the Prince of Wales might be assumed not to possess one – his own wireless receiving set. The status and reach of wireless had certainly come a long way over the intervening fourteen years. By 1936, over 30 million people in a population of 47 million were able to listen. Between these two famous broadcasts, though, there was a proposed third speech, which nobody heard because it was suppressed at the time. Its intent and contents show just how central broadcasting had become.

In this speech (released only in 2003 by the Public Record Office[3]), Edward had hoped to win public support for his marriage to the divorced American socialite Wallis Simpson. He had planned to begin with these words: 'By ancient custom, the King addresses his public utterances to his people', then continue: 'Tonight I am going to talk to you as my friends – British men and women wherever you may reside, within or without the Empire', and to explain how he hoped to marry and share a home with the woman he loved while remaining their king, trusting that his people would wish him 'to be blessed' with the same good fortune in happiness as they themselves.

Edward, though, was not like other people. Fearful of a constitutional crisis, the prime minster, Stanley Baldwin, had blocked the speech on the grounds that the sovereign could make no public statement on any matter of public interest except on the advice of his ministers. Denied the airwaves on 3 December, he had adapted some of these words and phrases and used them in the famous version which he then gave to the nation a week later, on the day after he had signed the declaration of abdication.

There is, perhaps, in the intimate tone and content of the undelivered address a little touch of Shakespeare's Henry V on the night before Agincourt. Edward also clearly already understood the guiding principle of radio presentation – that it should sound like one individual addressing another – and yet, in invoking the ancient custom of a king addressing his entire people, he had taken it wholly for granted that the medium through which he would do so was the wireless.

Only a year later, E.M. Delafield was to write her book, *Ladies and Gentlemen in Victorian Fiction*, in which she made her point that wireless had become the most significant symbol of modern home life. It didn't matter whether you liked jazz or loathed it, preferred classical music to variety, adored Henry Hall's dance-band music or were content to have it on in the background, nor what you considered your main source of news; Delafield's point was that wireless had now become ubiquitous. With this new accommodation of wireless in the home, the relationship between broadcasting and its listeners was once again changing.

Paradoxically, at the very time when listening to the wireless was assuming its 'part of the furniture' status, a new generation of designers was transforming the gadget itself so that it resembled the furniture only of the most fashionable and modern of homes. The 1930s' object of desire was a table-top or mantle radio which contained its electronics and speakers internally, was easier to control, could be moved from room to room and plugged into the mains. Above all, it was glamorous in design. The first on the market was made by the EKCO company. Eric Kirkham Cole and his future wife Muriel Bradshaw had been making radios in Southend-on-Sea in Essex since 1924. The company really took off after Cole read an article in the local newspaper by a schoolmaster and freelance journalist who wondered whether wireless sets might be powered from the mains. By 1930 the firm had moved to a spacious factory and begun manufacturing a new, non-flammable plastic which it had previously imported from Germany. 'Bakelite' was not only itself a symbol of the machine age: the fact that it was mouldable inspired a wholly new style of streamlined designs, many of which would not have been feasible in wood. It was EKCO that produced the famous round radio, the first of its kind in the world. Soon the company was commissioning young men with international backgrounds in architecture and industrial design, such as the Russian-born Serge Chermayeff and the Canadian-born Wells Coates.[4] Other companies followed suit: Philips brought out its starburst image; Pye's much-loved rising sun design seemed symbolic of a new dawn. In some ways, though, the greatest innovation of all was the minimalist wooden cabinet model dreamed up by the furniture designer R.D. Russell for Murphy.[5] It was said that some proud owners considered their clean, elegant lines so attractive that they went out and bought new furniture

EVERY GREAT STYLE HAS BEEN THE CREATION OF BEAUTY OUT OF USEFULNESS
IN THIS NEW YEAR OF 1934, RADIO LED BY PYE RECOGNISES A NEW RESPONSIBILITY BOTH ARTISTIC AND MUSICAL TOWARDS THE HOMES OF GT. BRITAIN
PYE CAMBRIDGE RADIO

Pye radio advertisements positioned the company's designs as both classic/classical and modern/modernist.

to match: a case no longer of wireless fitting into the home but the home adapting to the wireless set. Persuasive advertising copy in newspapers and sales brochures spoke to a range of tastes and pockets but was united in the message that these were new radios for a new age. Ironically, the colourful radios which were taking off in other countries did not gain real favour in Britain until the late 1940s: the most popular colour in the 1930s was still brown, though it was now a mahogany- or walnut-effect brown rather than real wood. Cautious customers seemed incapable of letting go of the idea that their new sets should blend in with their old furniture.

Although the Central Electricity Board had now been wiring up British homes for a full decade, poorer housing was often supplied with only two electric sockets. Some households simply connected their mains wireless to the central light socket rather than plugging it in at the wall, but others continued to use the heavy wet-cell accumulators, or batteries, which had to be recharged regularly at a local garage, cycle shop or radio supplier. The shopkeeper in *Dancing at Lughnasa* tells the sisters that they go through their batteries quicker than anyone in Ballybeg. It was certainly an inconvenience: children in working-class Warrington hated lugging them up the street, while some of their house-proud mothers worried that leaked acid would stain the carpet or strip the polish from the dresser.[6] At least now, though, all listeners could rely on easy controls, good-quality sound and fairly interference-free reception.

Whatever set they were tuning, most were soon looking forward to another major public occasion. As the turmoil of the abdication faded, plans which had already been under way for the coronation of Edward VIII were swiftly adapted so that his brother could be crowned as King George VI. It was just the kind of occasion which the BBC had made its own, combining a solemn national narrative with popular glamour. Now, though, a wholly new method for gathering public opinion and sentiment would shed fresh light on the listening experience.

The coronation took place on 12 May 1937. Every aspect of the day was recorded in a book edited by Humphrey Jennings and Charles Madge, two of the three founders of Mass Observation. This pioneering social research project had been launched after Edward's abdication, its aim to chronicle everyday life in Britain through diaries and questionnaires filled in by a panel of around five hundred untrained volunteers. To read *Mass-Observation Day-Survey: May 12 1937* is like joining onlookers at a promenade performance of a national drama, eavesdropping as it roams up and down the country, moving in and out of public buildings and private houses. Excitingly, this landmark book contains an extensive section on 'Reactions to Wireless'.

The three central events were the procession, the ceremony itself and the king's speech. When asked about the funniest incident of the day, one respondent recorded an elderly woman asking, 'Did they have a coronation in London as well?',[7] but there were few people now who did not understand how the wireless brought them together as a nation, regardless of where they were listening.

The coronation occasioned another return to shared listening in public places. A man who had rented an observation spot on the seventh floor of a block of flats at the corner of Marble Arch and Park Lane ('we sit in exceeding discomfort at the rate of 12 guineas each'[8]) commented on the street outside sounding like a vast radio shop, with 'the loudspeakers muttering most of the day and a small child in a pink coat asking "What are they shouting away for, sillies?"'[9]

A respondent in Southport listened with a group of pop-drinking boys in a cinema which had opened its doors for free, and was dismayed to report how badly they behaved.[10] Another described a packed pub in Ruislip where, above the din, a loudspeaker was blaring forth the ceremony from Westminster Abbey: 'At 12.30 came the words "The King is acclaimed". A rather noisy hail-fellow-well-met man in plus fours shouted "The King God bless him". People stood for a minute, then resumed their seats, continuing drinking, and apparently forgot there was a coronation.'[11]

A hearty group of physical-training students on a camping holiday in Sussex listened to the king's speech on their portable set, 'and there was an old boy with a row of medals coming along the road. So we asked him if he'd like to listen. He stood stiffly at the salute during God Save the King.'[12]

At home, some listened in silence, while others provided their own running commentaries. A woman in Scotland was concerned with the practicalities: what did the king and queen do for food all day and how did the troops lining the route relieve themselves?[13] A Nottingham hairdresser kept her wireless on continuously from half past ten to half past four: 'And you should have seen my mother – she sat in front of it all day – and all through the service while he was being crowned and that, the tears were pouring down her face and she kept moaning: "Oh, it ought to be Edward – it – it – it ought to be Edward".'[14]

Sympathy for Edward was not uncommon, and there were even rumours that he had secretly travelled from France to London and mingled with the crowds in disguise. But there was also kindness about the periodic hesitations caused by the new king's speech impediment. A woman in Swansea observed that many had noticed this ('You know, you'd think he'd finished and then he'd go on again'[15]), but that, in view of the strain of the day, 'he did very well'. A young woman in North Shields paints a vivid picture of a middle-class family home and describes how her reactions were altogether more powerful than she had anticipated:

> When we got home we found that Father was listening to the wireless, so we joined him with the Daily Telegraph supplement to enable us to follow the service intelligently. I did embroidery and mother mended while we listened. I found that my interest was decidedly quickened as the service proceeded; mother was more prepared than I was to be thrilled. I liked the music, especially the fanfares of trumpets, and as the ceremony proceeded I found myself surprisingly moved, until I felt that I wanted to cry. That certainly surprised me as I am not easily moved by plays or novels. It might have been the music or the profound solemnity and significance of the service.[16]

Across the country there were discussions about how to behave during the more sacred moments of the ceremony now that it was being beamed into individual homes. Was it permissible to eat during the broadcast? Was it better to focus exclusively on the wireless or could one also occupy oneself with knitting, sewing, reading, playing bridge or doing housework? How completely were listeners participating and how far should they behave as if they were actually present? People generally agreed that, wherever they were, it was right to stand up during the playing of the national anthem, but there was less certainty about other key moments, as one man reported from Hertford:

> Went back to the cricket ground for lunch. When 'The King' was played (the radio, which was on all the time in the pavilion) the men lunching on the balcony at once stood up, but those at the large table in the pavilion remained seated. Those outside noticed this, and rapped on the windows, and one put his head inside the door and said: 'Stand up, you chaps, The King'. Those inside then got up.[17]

Many, though, showed rather less reverence, as one correspondent, who had spent part of the day in a cafe in Devon, reflected:

> We listened in from 1.20 to 2.20. This was mostly from within the Abbey. While the prayer 'Wherefore with Angels and Archangels and all the Company of Heaven' was proceeding, there was much clamping about in the kitchen, and at the 'Sursum Corda' there was much bustling and banging. While the Archbishop was reciting passages in the Holy Communion – 'Hear what St. John saith' – a customer came into the outer shop and gave his order in the same ecclesiastical tones, for tobacco.[18]

Such responses partly reflected a shift in the relationship between subjects and monarch, congregation and church, but – however highly mediated – wireless itself was playing a role, as it had on the night of the old king's death, in bringing major public events into listeners' homes. Able to bear witness in this immersive way, listeners were unashamedly sharing their actual responses. The South Londoner who reported his party's comic misinterpretations of what had been said ('Gawd blimey' and 'swelp me bob') was reflecting a cultural shift in which the earlier sense of awe and wonder about wireless itself was being replaced with a less deferential attitude.

A crisis such as the abdication or a celebration like the coronation might focus listening for a few days, but the BBC was at last recognizing that it needed to understand its massive and diverse audience in a more regular and formal way. The previous system was simply not fit for purpose: the weekly postbag was huge, but opinions expressed in it were, by definition, self-selecting, while anecdotes gathered from the friends and acquaintances of BBC staff were, of course, shockingly unrepresentative. Infamously, a senior manager who had long argued that 'nobody dines before 8' had been surprised to discover that the vast majority of the population had already finished their evening meal by seven.[19] The significance of such knowledge in terms of attracting and keeping an audience was at last seeping through. In late 1936, just before the abdication, the first Listener Research Officer was appointed.

Robert Silvey was only 31 but he was clever and thoughtful, with a useful understanding of the commercial radio sector. Like his friend Winifred Gill, he was a Quaker, and had a good ear and an exceptional command of metaphorical language. There was much fear within the BBC, for example, about how his research might be used, so he emphasized from the start that his approach would be sociological rather than statistical, and that the role of his department was in 'map-making' rather than 'navigating'. In other words, he understood that programmes should not be made simply to satisfy the greatest number of listeners; nor should his research constrain adventures into as-yet unknown territories. His earliest work on the tastes and habits of listeners confirmed what some BBC staff had already supposed, that they were divided not along the lines of region or gender but of age and social class. It was partly because of work carried out in his unit that this divide was soon being breached by a range of programmes which were more inventive than anything listeners had ever heard before.

What characterized this new output was a thrilling new complicity between broadcasters and listeners. In the earliest days of broadcasting, the local stations in particular had spoken to their listeners with an easy informality that arose at least partly from a sense of mutual experience and background. The late 1930s saw its return. As Paddy Scannell and David Cardiff explore in the first volume of their seminal work, *A Social History of British Broadcasting*, this was built around a kind of listener participation. It drew heavily on Victorian parlour games, quizzes and old jokes but subverted them, spinning this tradition into a sophisticated and surreal style of comedy which was based on a shared understanding that radio might be more of a two-way street than had previously been supposed or allowed. On the one hand, there was an agreed joke that broadcasters could see into listeners' homes and hear them too, for example when the presenter of a series titled *Sing-Song* exclaimed, 'Now Granny, put your knitting down' and 'There's a woman in Hull not singing!'; on the other, a recognition that, while listeners clearly could *not* see into programme studios, they knew an absurd sound effect when they heard one. This blurred the boundaries between Broadcasting House and listeners in their own homes – whoever they were, whether they lived in a mansion or a cottage, a suburban semi or an urban back-to-back. As Scannell and Cardiff put it: 'Like the private jokes of a family, radio comedy built up its far-fetched associations, while remaining rooted in the charm of the familiar. One resource that could always be relied on as a shared point of reference available to all listeners was the culture of radio itself.'[20] After a fifteen-year gestation period, this new, radiogenic brand of comedy was as unique as a newborn baby. Its parents, grandparents and godparents were rooted in the past but it belonged to the present and to the future.

The first and perhaps greatest showcase of radiogenic comedy was entitled *Band Waggon*. Because listener research was revealing that dance music alone was not as popular as had originally been thought, *Band Waggon* was conceived as a compendium programme featuring comedy alongside music. Soon, though, it had a duo of uncontested stars. Like Mr Murgatroyd and Mr Winterbottom, they were – significantly – a socially mismatched but brilliantly sparky duo: Richard 'Stinker' Murdoch, who had been educated at Charterhouse before finessing his comedy talents with the Cambridge Footlights, and the diminutive Arthur 'Big-Hearted' Askey, who had worked as a clerk in Liverpool and gained experience in seaside concert parties. The first few episodes flopped, and the series was almost pulled. Then Arthur Askey began to riff on the idea that, because he was billed as the 'resident comedian', he actually resided in Broadcasting House, and that a BBC announcer would come round every night after the show to tuck him up in bed and bring him a glass of milk.

Along with their production team, the two comedians quickly developed a fabulously rich and surreal fantasy life which soon caught on with listeners: they lived in a shared flat on the roof of Broadcasting House with a goat called Lewis and four pigeons called Basil, Lucy,

After the massive success of *Band Waggon* on the wireless, Richard Murdoch and Arthur Askey starred in a 1940 film of the same name, which featured an overpersuasive estate agent, a ghost, Nazi agents, police officers and a senior BBC figure.

Ronald and Sarah; they had a charlady called Mrs Bagwash, who never spoke; her daughter, Nausea, whom Askey was understood to be courting, was also entirely silent because she fainted whenever she approached the microphone.

As with other new output, *Band Waggon* was a series, broadcast in the same slot each week in order to build up audience loyalty. This created an opportunity for iterative jokes and catchphrases: 'Hello, playmates', 'Ah happy days' and 'Light the blue touchpaper and retire immediately', which was used in anticipation of any explosion by Mrs Bagwash. Catchphrases had been used before, but these took off as never before, and one in particular was the first to enter the language: 'Ay thang yew', a strangled expression of exaggerated gratitude, picked up by Askey from a London bus conductor, was soon to be heard across the country. Like all the best comedians, the two sounded as though they were speaking off the cuff: as Askey commented to Murdoch, 'the trouble with you is you know so little and know it so fluently'. The shotgun speed of their repartee – a sort of accelerated, amplified version of everyday conversation – was also a novelty as the majority of broadcasters had, until recently, spoken in such a slow and often solemn and rather stilted manner. Listeners from all backgrounds could easily pick up and repeat their lines, integrate their catchphrases and generally enjoy a sense of belonging to a new club. They could also share in Murdoch and Askey's affectionate mockery of 'the dear old BBC'. If the pips, the shipping forecast and the SOS announcements had launched the project to unite listeners in a sense of shared nationhood, programmes such as *Band Waggon* brought them together through laughter.

Very little audio survives of the fifty-five episodes, but you have only to listen to the skit in which Askey and Murdoch purportedly sit down side by side at the Broadcasting House organ (normally played by the BBC's resident organist Reginald Foort) to understand why the series soon became a cult: first, with popping sound effects, they pull out all the stops; then, like two children at a painfully out-of-tune household piano, they thump out an ecclesiastical version of the tune known as 'Chopsticks'; finally, Askey accidently presses a button which – as if he is now sitting at a cinema organ – sends him up and through the roof of Broadcasting House before dropping him down, with cries of 'Help, mother, I haven't got me parachute!' against an exaggerated clatter of sound effects. So large was the audience of *Band Waggon* that it soon overtook in popularity the long-running Saturday-night variety favourite *Music Hall* (its Victorian title increasingly outmoded). Alternative forms of entertainment also keenly felt the competition: during its runs, cinema and theatre bookings dropped on Wednesday evenings, and the Post Office reported a reduction in the number of telephone calls.

There were a (very) few who didn't appreciate the new kind of comedy. Mrs Evans confided her rather poignant opinion to Winifred Gill in Bristol:

> I don't like the wireless. I'm jealous of it. I used to keep the 'hole 'ouse in roars of laughter, sayin' daft things. We was all like that. When we got the wireless, it was 'old yer 'ush we want to listen to this!' When I'm alone, I carries on like mad – I don't know what the neighbours think – I fair scream the 'ouse down. I think I was made to be 'appy. No, I think the wireless as taken 'alf the life our o' me. [21]

But Mrs Meyer's opinion was closer to that of the vast majority of listeners, across region and class, age and gender: 'I often repeat the jokes myself to my husband. Really, I don't know what we should do without it', she told Gill, before turning to her husband and repeating these final words: 'Really, I don't know what we should do without it.'[22]

⁂

Radio – along with the experimental new medium called television – was a natural vehicle for conveying the coronation to the people. Once the technical hurdles had been overcome, the BBC's success in sharing such major national occasions was assured. Its triumph in this new kind of comedy, on the other hand, was more of a surprise, and contains a lesson which is as important today as in the late 1930s.

The commercial stations had built up their popularity by being attentive to public demand. Listeners already knew that they loved light music and variety, so Radio Luxembourg and Radio Normandie gave them these in abundance. Reith, meanwhile, had stuck to his guns and his high-handed early justification for 'apparently setting out to give the public what we think they need – and not what they want' on the grounds that 'few know what they want and very few what they need'.[23] The clear implication was that he and his team knew best what would be good for the public, as they were ignorant both of what this might be and of their own true taste and appetite. In fact, in the early 1920s when wireless was a brand-new medium, even the broadcasters had simply been feeling their way in the dark. All they could do was plunder pre-existing art forms and draw on their own instincts to discover what would or would not succeed. Since they themselves did not really know what they wanted or needed, it was hardly surprising that this was also true of the listeners. By the late 1930s Robert Silvey and his research unit might collect and share information about public responses to on-air output, but, as he insisted, they would do no more than provide a map of the territory to be navigated. Even with a captain at the helm as autocratic as John Reith, it was the creative crews who would have to be trusted to use these maps, to explore their furthest reaches and even venture beyond. The public could then decide whether they wanted or needed their new discoveries. In the case of comedy, it turned out that they wanted them very much indeed. Reith's greatest leadership quality, by nature, was not the ability to release or enable creative ambition in others, but there lay hidden within his words – hidden even from his own understanding – a supremely important manifesto: listeners do not necessarily know what programmes they want until creative teams develop and produce them. On some occasions these teams will fail; on others the listeners will so much love what they hear that they will soon consider it incredible that they did not know about it all along. In this, the listeners will themselves have been active participants, collaborating in a creative journey which pushes back the boundaries of the uncharted airwaves.

In a desperately alarming world, the benign madness of radio-genic comedy was soon to give listeners not only pleasure but also reassurance. At the same time, with recognition growing about the mobilization of radio in other countries for malign ends, the BBC's provision of news was to confirm its place in the trust and affection of the British listening public.

RADIO TIMES, NOVEMBER 18th, 1938

RADIO TIMES

HOME NUMBER

TEN

THE HOME SERVICE

> Stands to reason. It's more direct, more official ... [they] wouldn't tell lies out loud for everyone to hear.[1]
>
> Bristol housewife

When prime minister Neville Chamberlain touched down at Heston Aerodrome to the west of London on Friday 30 September 1938, he was returning from what had been his third trip to Germany in only two weeks. A few days previously he had explained the reason for this desperate act of shuttle diplomacy in a broadcast from Downing Street which was transmitted both to the nation and to the empire. In this he expressed his horror and disbelief 'that we should be digging trenches and trying on gas masks here because of a quarrel in a faraway country between people of whom we know nothing...' Now, standing on the rainy runway, he held aloft one of the most infamous sheets of paper in history and read its contents to the cheering crowd:

> We, the German Führer and Chancellor, and the British Prime Minister, have had a further meeting today and are agreed in recognizing that the question of Anglo-German relations is of the first importance for our two countries and for Europe. We regard the agreement signed last night and the Anglo-German Naval Agreement as symbolic of the desire of our two peoples never to go to war with one another again.

Contrasting a decade and a half of sunnily optimistic *Radio Times* covers, this sombre November 1938 image, by John Rowland Barker, communicates the sense of threat increasingly felt by broadcasters and listeners alike.

Far from bringing a sense of relief, Neville Chamberlain's famous press conference at Heston seems to have increased anxiety among reporters.

Chamberlain's day was not over. To enthusiastic shouts of 'Three cheers for Neville!', he was driven away to Buckingham Palace, where, alongside his wife Anne and the king and queen, he appeared on the balcony before even larger crowds, once again expressing their relief that the Munich Agreement had apparently brought the threat of war to an end. Next, in front of 10 Downing Street, he repeated the statement he had made at Heston but with additions which, in retrospect, can be read only with a sense of the bitterest irony. Echoing a phrase of Benjamin Disraeli on his return from an 1878 diplomatic conference, Chamberlain declared: 'My good friends, for the second time in our history, a British Prime Minister has returned from Germany bringing peace with honour. I believe it is peace for our time.' Finally, no doubt longing by now for his own bed, he concluded with this reassuring suggestion: 'Go home and get a nice quiet sleep.'

Chamberlain had become a Member of Parliament in the 1918 general election, which had immediately followed the armistice. He was was far from alone in his forlorn wish to avert the slaughter of another world war. The first phase of the history of wireless falls almost precisely into this fragile era between the end of one world war and the beginning of another. For the BBC, the image which the prime minister had conjured of home as a haven safe from the horrors of war was equally sacred. Over the last seventeen years it had been in the home that wireless had become firmly established. Before long it would be in this space that listeners would come to understand that Munich had failed and begin to learn a new way of living as citizens of a country at war.

⁂

During the earliest days, prominent articles in the *Radio Times* had celebrated the ways in which wireless was providing 'a new call to the fireside' and ushering in 'the revival of home life'. These trod a narrow path between disdaining and praising the Victorian parlour and all it represented. On the one hand, they optimistically suggested that wireless was 'rediscovering the charm and happiness of home life, and by building happy homes it is welding the love that will hold the Empire together through sunshine and shadow, through peace and war'.[2] On the other, they agreed that most Victorian domestic amusements had by now very much lost their appeal, particularly to young men. Magic lanterns, billiards tables, ping-pong, jigsaw puzzles, the gramophone, spelling bees, recitations, reading aloud, and, in Scottish homes, memorizing the Shorter Catechism – none of these could compete with pubs, nightclubs or the pictures. Wireless, by contrast, could make each home a place of entertainment and truly appeal across the generations.

The 1925 all-colour Christmas number of the *Radio Times* tells the story through a single image: in the foreground, a pair of young women in slinky silk slip dresses – dazzling shades of green and orange – dance with two cigarette-smoking young men in informal suits to light music on the wireless. In the background, the grandmother and

The *Radio Times* cover from Christmas 1925, at a time when broadcasting was often still baffling to the elderly though enthusiastically embraced by younger people.

grandfather – their clothes and hairstyles distinctly grey and Victorian – observe the party. They are bemused by the scene before them but happy to observe the young people enjoying themselves at home.

If in the mid-1920s, as this image implies, the duller aspects of Victorian domestic life and the dubious attractions of external entertainment were still seen as potential rivals to the attractions of home, a decade later it was already becoming apparent that the enemy of home was the German war machine. From 1935 onwards a sequence of special *Radio Times* 'Home' and 'Fireside' numbers reinforced this idea.

Only seven weeks after the ostensibly triumphant resolution of the Munich crisis, the front cover of the *Radio Times* was stark in its depiction of a country under threat, or, rather, a single home in peril. Drawn by the artist John Rowland Barker (who sometimes used the pen name Kraber), it depicts an isolated white house set against dark hills at night. A brick path leads from the garden gate to an open door. The lights inside are blazing but the curtains have not been drawn and the house seems strangely empty, exposed and vulnerable. Meanwhile, on either side of the garden stand two almost leafless trees, a stunted white one in the background and, in the foreground, a massive black one which towers menacingly over the pitched roof and dominates the entire image. The lettering declares 'Home Number', but there is nothing cosy here. The editorial copy is similarly sombre in tone. In his regular column, W.H. Elliott, the BBC's Radio Chaplain, refers to the recent crisis, while, in an article re-evaluating the music of Mendelssohn, the composer, conductor and writer Constant Lambert is brutally frank about the situation in Germany.[3]

Between the front cover and the articles, though, there is an inner title page which features one of the most charming and at the same time strikingly eloquent cartoons in all the 850-odd pre-war editions of the *Radio Times*. In scratchy black pen and ink, Mervyn Wilson has drawn an image which brilliantly blends the now familiar lines of Broadcasting House with architectural details of a cosy suburban home and an idyllic country cottage. There is a picket fence, hens cluck in the yard, and washing hangs over the flourishing kitchen garden – towels

This image of Broadcasting House by Mervyn Wilson brilliantly blends the home of radio with a listener's idealized home. It was published in the *Radio Times* less than a year before the outbreak of the Second World War.

and underwear pegged to a line which stretches from a pole to the walls of the building itself. A comfortably plump man in suit and hat, briefcase in hand, strides happily home to his wife, who stands waiting under a pergola at the gate. Above and behind her, like benign sentinels, stand the statues of Prospero and Ariel, who guard the famous bronze doors leading into the double-height reception of Broadcasting House. Softening the curved art deco walls, a wisteria in full bloom climbs up towards the iconic clock and transmitter, the branches making their way between numerous windows which are each dressed with spotted curtains. It is both an unashamedly romantic image and a poignant representation of everything which war threatens. The text reminds readers that 'when all is said and done, broadcasting, with all its elaborate mechanism, is based on and aimed at the home'. It goes on to remind them that this is the unique quality of radio, pointing out that if wireless had been consumed as theatre and cinema are – outside the home – it might have been grander, more imposing and sophisticated, but altogether less intimate and friendly. And it concludes by pointing out 'there are homes on both sides of the microphone, and broadcasting is the connecting link between them, even if there is not as yet a wisteria-covered porch in front of Broadcasting House!' This, it implied, was the British life threatened by war.

⁂

By 1938 almost 34 million people in a population of 47.5 million were able to listen to radio. Even in a poor area such as Barton Hill, one male interviewee could casually remark that there were precious few[4] who did not now own a wireless. While interviewing small groups with particular interests, Winifred Gill kept in her notebook a useful tally, which is skewed towards women but vivid in its specificity: within what she called the 'Girls' Group' all seven interviewees had access to wireless; among the 'Garden Lovers' seven out of eight; in the 'Rug Class' ten out of eleven; among the Sunday-school teachers five out of seven; in a larger group which she called 'Mrs Gould's Class' twenty-two out of twenty-six.[5] Fewer and fewer now needed to feel themselves excluded, even among those who did not have – or could not rely on having – a place to call home at all. Mrs Hennessy, the wife of a local Labour politician, shared an experience, which she herself had clearly found moving, of a group of foster children who, at just this time, were taken to the seaside for a summer holiday:

> Mrs Lamb brought a beautiful wireless, and do you know, we couldn't get the children away from it. You'd have thought they'd have wanted to run down to the sands, but no – there they were, boys of 11 and 12, sitting round, their heads all close to the wireless, and they said, 'oh come here, Mrs Hennessy, the music's lovely'. They don't get much music in the homes...[6]

She went on to report that, at the end of the holiday, the children were given a wireless set so that in the foster home where they lived – ten

A LOST ART.
THE DINNER-PARTIES OF OUR ANCESTORS WERE EMBELLISHED WITH SPARKLING CONVERSATION—

This December 1923 cartoon predicts the success of wireless and prefigures with uncanny accuracy discussions in many twenty-first-century homes about the impact of mobile phones at the dinner table.

children to one mother – they could continue to enjoy listening like almost everyone else around them.

As the sense of crisis grew, though, wireless was increasingly valued by listeners as very much more than a source of entertainment or distraction. One woman described this increasingly critical role in the simplest of terms – 'You can't go far from the wireless'[7] – while Mrs Barnes, who hadn't listened since the General Strike (and then, presumably, in public), acknowledged that 'our Ken and Eunice keeps on at me and doubtless they'll soon get me persuaded'.[8] Mrs Ball, who voiced her personal dislike of the medium in one of Gill's group sessions, was met with a chorus of disbelief:

> Mrs Ball: I never turns it on. I don't like the wireless. I puts up with it when the children have it on.
> Mrs Keep: What never? Not for the news?
> Mrs Ball: No.
> Mrs Privett: Then if we was at war, you wouldn't know it, not till the Germans came marching up the street.
> Mrs B: COURSE I should![9]

BUT NOWADAYS THE TALKING IS DONE "OFF."

Compared with twenty years previously, listeners' interest in current affairs had become keener and knowledge of recent happenings at home and abroad far wider spread. Households surveyed by Rowntree in York consistently mentioned listening to the *News*. Gill and Jennings were keen to emphasize that this was true not only of women and men but even of what they call older schoolchildren – they were actually aged between 11 and 14. By chance, the questionnaires, which 841 children filled in, had been issued to them on 13 October 1938, only a fortnight after Chamberlain's return from Munich. Although they were not specifically asked about news at all, more than one in ten mentioned it of their own accord in reply to a question about the most interesting things they had ever heard on the wireless. Of course, this was partly because it was fresh in their minds and clearly a common topic of conversation among adults, but their references to 'When Mr Chamberlain went out to Germany to make peace' and 'The crisis about the war' suggest that the growing generation increasingly turned to wireless when trying to make sense of the world around them. A schoolteacher of many years' experience confirmed this, and the leader of a club for boys over the school leaving age was more expansive:

> The child of today is more of a general conversationalist than he was before the era of broadcasting, and his knowledge of topical subjects such as the international situation, though perhaps rather vague, is mostly obtained through the wireless, if not through actual listening himself, through listening to his family discussing broadcasts.[10]

This habit of turning on the wireless to keep abreast of news had so quickly become commonplace that it is easy to forget what a novelty it was.

⁂

Back in November 1922, 2LO had launched itself onto the air with the voice of Arthur Burrows reading the *News*, but there had then followed the long period when only a single bulletin each evening was permitted – in other words, news could only be broadcast once the newspapers had been bought and read. It was also agreed that the BBC would not gather news itself. In fact, the station identification, which was later to become such a familiar introduction to international broadcasts during the Second World War, had, in its original form, been followed by an on-air plug for the wire services which provided it: 'This is London calling – 2LO calling. Here is the first general news bulletin, copyright Reuters Press Association, Exchange Telegraph and Central News.'

The lack of newspapers during the General Strike of 1926 had led to a relaxing of the rules concerning news, with five bulletins broadcast across the day, but a return to the old schedule was soon enforced. Although the BBC had gradually gained the right to edit is own bulletins, it was not until 1934 that the *News* section had become separate from talks, and – astonishingly – it was only at the onset of the Second World War that news could be broadcast before six in the evening. In 1938 memories were still fresh in some listeners' minds of how important broadcasting had been during the General Strike. There was a widespread – though not by any means universal – agreement that BBC *News* had helped suppress rumour, reinforce morale and – most importantly – provide reliable information. With this current crisis, there was no shortage of newspaper coverage, but there was by now a marked preference for news provided on the wireless. One of Jennings and Gill's special-interest groups consisted of local newsagents, and Mrs Britton offered this lively analysis: 'The Herald and the Mirror are the most scandalizing papers there are, and people who like scandal buy them. The Sketch and the Express are different, and the Chronicle will even belittle a situation till they have heard the broadcast.'[11] Against her own commercial interests, she went on to clarify her personal preference: 'I seldom read the papers. I've no head for details, but the papers always exaggerate to get a sale. If you want the actual truth, you must go to the wireless.' Of the twenty-one younger women attending the Mothers' Group, all expressed a preference for wireless news over the newspapers, and the vast majority of Barton Hill interviewees agreed.[12] This was perhaps grounded in the BBC's carefully curated reputation for seriousness and its association with other trusted public institutions such as the church and the monarchy. Many people were certainly inclined to speak of it in a surprisingly deferential manner. Miss Vile declared that 'when it comes to news – different papers tell different things, conflicting stories. Wireless has authenticated responsibility behind them.'[13] Another woman concurred: 'Well they wouldn't say what wasn't true. It's just facts. You believe what's broadcast.'[14]

Across society, the BBC was now increasingly considered the superior source of news. A 1938 *Punch* cartoon shows a prosperous couple sitting by their fireside surrounded by recently read and now discarded newspapers; the man glances up at the clock and says to his wife 'We've missed the News, Edith.'[15] In Glasgow, Miss French, a single, middle-class woman working in the offices of a shipping company and living with her mother, responded to a Mass Observation directive during the weekend before Chamberlain's return from Munich: she reported that, apart from on Sundays, when she read the *Observer* and the *Sunday Times*, she actively avoided the newspapers, afraid that reading or even looking at their illustrations would ruffle her calm: 'My knowledge of the crisis comes from the wireless, for I never miss any news bulletins that I can possibly have.'[16] A few days before war was declared, Virginia Woolf, dazed and depressed, wrote in her diary: 'Yes, we are in the very thick of it. Are we at war? At one, I'm going to listen in.' Then, in the first full week of war, she observed for herself what the newspapers had long feared: 'All meaning has run out of everything. Scarcely worth reading papers. The BBC gives any news the day before.'[17]

Newspaper sales in fact remained buoyant. Some clearly agreed with Mr Pope that 'If you want to know about a thing, you must READ about it, and read it more than once';[18] others with Mrs Britton's observation that 'People don't buy papers for the news. They buys them for the sports, competitions and cross words.'[19] It was just at this time that the newspapers were undergoing a major makeover. Mark Pegg's meticulous analysis of the first two decades of broadcasting has provided the statistical underpinning for this and many other books about early radio, and he outlines some of the reasons for these changes. Anyone who has ever browsed through an early-twentieth-century newspaper, or perused one online, will be familiar with the eye strain caused by their crammed pages. Gradually now, readers found there was less fine print, more bold and large headlines, double-width columns and photographs. Some of these developments were brought about by advances in technology and because cinema newsreels were inspiring more visual approaches to telling news stories, but it was largely in response to the competition from radio that the newspapers of 1938 were beginning to broaden their function and to look very much like ours do today.[20]

Some people found any kind of reading hard, but there were also those who felt that the wireless helped them make better sense of the complicated and frightening international situation. One woman simply said: 'What's said to you is easier to take in than what's read.'[21] Mrs Morse relished her growing understanding and shared it enthusiastically and at some length. At the time of the German invasion of Czechoslovakia, listening to the BBC had helped educate her at a basic level:

> Pronunciation is a great help. 'Sudeten'. I first called them the 'Suden Germans', but after a while, I found I could say 'Sudeten' if I tried.

Then, as her confidence grew, she found herself able to explain matters to others:

> I've a neighbour who's ever so weak, p'raps not stupid, but she can't understand the wireless. When the crisis was on, she kept on coming in to see me and saying 'has anything happened? Is there any news?' And I'd say, 'but you've heard the wireless?' She'd say: 'Yes, but I can't understand what it's all about. But if you explain what's happened, I can understand.' One day about five she calls over the fence 'Who did you say Hitler had told to mind their own business?' And the names she suggested made me laugh – they wasn't any of them anything to do with politics. So I says, 'Eden, Duff Cooper', and she says 'Excuse me a moment', she said, 'I'll just go and tell my son before I forget.' Duff Cooper she might have let slip, but you'd think that ANYONE 'd 'ave remembered Eden![22]

For her, wireless news had truly become a reliable source of information and intelligence in a bewilderingly complex world: 'Wireless has no axe to grind – no political views. Papers have to be one party, with one point of view – the wireless tries to get all points of view.'

There was a small, articulate minority of Barton Hill residents, all of them men and all politically engaged, who disagreed with Mrs Morse's endorsement. What they seem to have been expressing was an understanding that reliability was not the same as impartiality. In this they were voicing the suspicious attitude towards the BBC long held by some working-class and unemployed people and others from the middle classes. This stretched back at least to the General Strike, when the BBC had certainly shared reliable information but in reinforcing the government side over that of the strikers had not fairly told all sides of the story. Lessons had been learned since then, though it's important to note that even on Friday 30 September 1938 the BBC was far from entirely independent: while it had assiduously provided live coverage of Chamberlain's arrival at Heston, appearance on the balcony of Buckingham Palace and speech to a crowd of five thousand outside Downing Street, it had suppressed information about a demonstration being staged just down the road in Trafalgar Square in which over three times this number had taken part.[23]

Once again the broadcaster had been following a government request. In Barton Hill, Councillor Rogers was one of the men who recognized issues such as this: he acknowledged that the wireless had generally had a positive effect in providing political education but was clear that 'all news and talks are so heavily biased that in fact it is all very reprehensible', and went on to remark that 'It's not till I took to listening that I became really class-conscious.'[24] Labour Alderman Hennessy agreed that there had been a general awakening of interest in politics and put this down to four factors – the press, political parties, the Left Book Club and the wireless. His view was that it was 'generally recognised in factory and workshop that BBC talks are not impartial... The scales come down heavily on the side of one party [which] makes people suspicious of the sincerity of the BBC and of the content and principles of the social matter broadcast from time to time.' But he too concurred with the commonly held view that the corporation had

> rendered a service in bringing foreign countries into our homes. However much knowledge listeners may have gained, there is always something that only the BBC can supply – knowledge of the institutions and customs of foreign countries. The BBC can be of real help in creating interest in international affairs by supplying the human and institutional side.[25]

Knowledge and interest were only part of the package. Simply at an emotional level, there was reassurance in listening to the wireless. Mrs Pugsley was not alone in finding comfort and 'a calming, quietening effect' in the tone of the announcers' voices, even if they were simply reading the weather or an SOS call. Most importantly, she spoke for tens of millions when she simply said: 'In terms of news, I switches on to know whether we are at war yet.'[26]

In an unexpected way, at the very time when even the poorest people had acquired their own radios, the national crisis was reinventing an older, collective form of listening. Outdoor loudspeakers still featured in common memory: older people might recall marvelling at public displays of wireless as amusing gimmicks at bazaars in the early 1920s; many could think back to the nine days when they had huddled outside radio shops during the General Strike. But an editorial in the edition of the *Spectator* of 15 September 1939 (just after the beginning of war) was alert to a new version of this old experience: 'Whereas press news is imbibed singly, radio news is imbibed in groups, you can walk down a street and hear the same voice busy in every house. Thus radio news is community news: it is a united gesture of a society listening at the same time.'[27]

Of course the idea of the BBC as a shared institution had been growing in the public imagination since its inception, but its role as provider of wartime news was to develop this beyond recognition. In the same week as the *Spectator* article appeared, but in the (semi-fictional) rural Devon world of E.M. Delafield, an anecdote rings true about six evacuees from Bethnal Green and the redoubtable Miss Pankerton:

> Extraordinary legend is current that she has taught them to sing: 'Under a spreading chestnut-tree, the village smithy stands', and that they roar it in chorus with great docility in her presence, but have a version of their own which she has accidently overheard from the bathroom and that this runs:
>
> 'Under a spreading chestnut-tree
> Stands the bloody A.R.P
> So says the ——ing BBC.'
>
> Aunt Blanche, in telling me this, adds that: 'It's really wonderful, considering the eldest is only seven years old.' Surely a comment of rather singular leniency?[28]

British listeners, of course, had their counterparts all over Europe, each one turning on their own radio to catch up with the *News* and 'to know whether we are at war yet'. The difference lay in the wider intent of the broadcasts that they heard. In Germany, after Joseph Goebbels had taken up the role of Minister of Propaganda in 1933, radio had become an essential tool of the Nazi regime. Goebbels had seen to it that no listener would be excluded, arranging for a consortium of manufacturers to mass-produce a set which was comfortably within the budget of all German families. The tuning options of the austere *Deutscher Kleinempfänger* were limited. This was partly in order to keep the price down, but largely because those in authority wanted listeners only to hear the official story promulgated by *Deutschlandsender* and the local *Reichssender*. It's not just a judgement born of hindsight that this square black radio was ugly in design (its Italian equivalent – setting aside the fascist symbol built into the fretwork – was rather elegant); nor to suggest that the large circular mesh amplifier seemed to symbolize the transmission of a single message spoken by one man to 80 million Germans.

With their more powerful sets, many British listeners were able to pick up English-language broadcasts from Germany,[29] just as they had long tuned into Radio Normandie, Radio Luxembourg and the other commercial Continental stations. One piece of BBC research conducted in May 1939 asked listeners about the regular broadcasts from Hamburg and Cologne each evening. About a third of respondents acknowledged that they had heard them, but few listened – or admitted to listening – regularly. As one man put it: 'I felt, what is the use of all this, it's only a pack of lies – we can only get the truth from our own stations, and of course did not listen again.'

In contrast to such output, the BBC continued at least to try to share more than a single message – to broadcast a range of voices and viewpoints. The warning on Marconi's swordstick was proving prescient: NON TI FIDAR DI ME SE IL COR TI MANCA. Radio could be a dangerous weapon and this was no time for the faint of heart.

Entertainment was essential for keeping up morale, and anyone switching on the wireless was likely to catch a stream of what we would now call 'easy listening'. On the day Chamberlain returned from Munich, listeners could hear an organ recital, old favourites Henry Hall and his Orchestra, a commentary on a ploughing competition at Moreton-in-Marsh, a range of 'Rhythm Classics', the Victor Olof Sextet, Vita Sackville-West discussing famous gardens, music from Section C of the BBC Orchestra, a programme entitled *BBC Ballroom* which promised 'Dancing tonight to the music of Stanley Barnett and his Band, Admission by radio only...', a programme simply titled *Restful Music*, an 'Audible picture of life in the Royal Airforce' featuring Airman 47398 (the improbably named Harry

Hawk), a range of exotic music performed by Alfredo Campoli and his Salon Orchestra, Lew Stone and his Band, and further dance music on gramophone records right through until close-down. But alongside such output, the editorial staff of BBC News were enlarging their role, arguing that morale could best be maintained by treating the listeners like adults and giving them as truthful an idea as possible of how things were going, even when things were going very badly indeed. The writer Penelope Fitzgerald worked in Broadcasting House in her early twenties, just after the war had begun, though she published her magnificent novel *Human Voices* some four decades later. Her memorable description of the corporation as 'a cross between a civil service, a powerful moral force, and an amateur theatrical company that wasn't too sure where next week's money was coming from'[30] has never been bettered. Nor has that of its wartime mission:

> Broadcasting House was in fact dedicated to the strangest project of the war, or of any war, that is, telling the truth. Without prompting, the BBC had decided that truth was more important than consolation, and in the long run would be more effective. And yet there was no guarantee of this. Truth ensures trust, but not victory, or even happiness. But the BBC had clung tenaciously to its first notion, droning quietly on, at intervals from dawn to midnight, telling, as far as possible, exactly what happened. An idea so unfamiliar was bound to upset many of the other authorities, but they had got used to it little by little, and the listeners had always expected it.[31]

The mutual expectations of listener and broadcaster were at the heart of the British experience when war was declared just under a year after Chamberlain's return from Munich.

⁂

The power of radio and the key role it would play across the globe were apparent from the very beginning of the conflict, at both a practical and a symbolic level. On the night of 31 August 1939 a group of German SS officers disguised in Polish army uniforms attacked the radio transmitter at Sender Gleiwitz in what is now the city of Gliwice in Poland but was then in Germany. This was the pretext for Hitler to broadcast a speech condemning what had in fact been a false-flag incident. On the following morning German troops invaded Poland, and on 3 September Chamberlain made his broadcast from Downing Street, sorrowfully declaring that Britain was now at war with Germany. Listeners then heard a series of announcements informing them that all places of entertainment were to be closed with immediate effect and that any gatherings of crowds, except in churches, were to be avoided. At this stage, the expected bombs did not in fact fall, but with theatres, cinemas, concert halls and dance venues forced to go dark, the role of radio was becoming more important than ever.

From 1 September BBC output had been streamlined: the two networks, National and Regional, which had been in existence in most areas since the beginning of the 1930s, were now merged into one single network called the Home Service. This designation had already been in use to differentiate domestic output from the overseas service or the fledgling television service, but the name now had a new symbolism: it represented everything that war threatened. On the Home Service listeners could now hear a *News* bulletin 'every hour, on the hour', general information and a stripped back programme of mostly light music. This included an astonishing number of hours featuring live performances by the organist Sandy McPherson, who stood by to fill any gaps in the schedule. In fact, plans were soon hastily being made for an entirely new network, and in January 1940 the BBC launched its Forces Programme.[32] Initially this was intended to entertain British soldiers in France, but once listeners discovered that they could actually pick it up at home, it quickly became the go-to network for precisely the kind of programming which many had long desired: drama, comedy, popular music, quiz shows and variety as well as news and talks. On a day-to-day basis, there were soon far more listeners to the Forces Programme – particularly women at home and working in factories – than to the output which was intended for them. But it was to the Home Service that they turned at critical moments throughout the war. In particular, listeners across the country took upon themselves what became an almost sacred obligation to tune into the *News* bulletin at nine o'clock every evening.

The day after war was declared, the BBC had also rushed out a revised edition of the *Radio Times.* The photograph on the cover was of Broadcasting House, a Union flag flying on its flagstaff. Beneath the image was the declaration 'Broadcasting Carries On!' There was no wisteria in sight, no cheerful couple, no knickers and long johns hanging from the washing line, but the powerful message behind Mervyn Wilson's drawing took a deep hold in real life: that the home of the broadcaster and of the listener were in some sense one and the same. Once again, with her emotional honesty and dry wit, Penelope Fitzgerald sums up the insiders' experience of the BBC. Referring back to the origins of the word 'broadcasting' – now so entirely familiar that it was easy to forget it had been in use for less than two decades – she fondly recalls the day-to-day mood among those who were 'scattering human voices':

> everyone who worked there, bitterly complaining about the short-sightedness of their colleagues, the vanity of the news readers, the remoteness of the Controllers and the restrictive nature of the canteen's one teaspoon, felt a certain pride which they had no way to express, either then or since.[33]

With a similar mix of ambivalence and courage, listeners were just beginning to adapt to the new way of life. Six weeks after war was declared, a photograph on the cover of the *Radio Times* showed a group of miners, 'somewhere in Wales'. Further details could be found on page 22: preceded by a 'Topical Revue' billed as 'For Amusement Only' and followed by the now reassuringly familiar Greenwich Time Signal and the *News*, the programme was entitled *The Home Front.*[34] Its subtitle, *Home Fires Burning*, intentionally cast listeners' minds back to the song which had been so popular during the First World War. Once again, Britain was at war, but this time there was wireless.

for the radio epicure

GECoPHONE

REGISTERED TRADE MARK

RADIO RECEIVERS, LOUD SPEAKERS & ACCESSORIES

MADE IN ENGLAND

THE NEW "OSRAM MUSIC MAGNET" 3-valve home constructor's circuit. The sensational success of the season. Price complete with **OSRAM Valves,** solid oak cabinet, and full-size Instruction Chart £9

B.C.3045/7. GECoPHONE Portable Screen Grid 4-Valve Receiver. Maroon or brown leather finish. Price complete with **OSRAM Valves.** 23 gns.

B.C.3038. GECoPHONE All-Electric 3-valve Portable Receiver, in solid mahogany. Price complete with **OSRAM Valves** £28

B.C.1790/2. GECoPHONE "Stork" Cone Loud Speaker. Cabinet model in mahogany. Price £4
In oak £3 15 0

B.C.1770. GECoPHONE "Stork" Cone Loud Speaker. Senior Plaque model. Price 50/-

GEEKO L.T. and H.T. Wireless Accumulators. Will give years of lasting service and satisfaction. Special solidified type for Portable Receivers

B.C.3030R/1R. GECoPHONE 3-valve All-Electric Receiver in solid mahogany or oak. Price complete with **OSRAM Valves** £25

B.C.3020/25. GECoPHONE 2-valve All-Electric Receiver in solid mahogany or oak. Price complete with **OSRAM Valves** £15

MAGNET Wireless Batteries. Specially made for modern high-power receivers. Ensure economical and efficient operation.

Selections from the famous range of **GECoPHONE Radio Receivers, Loud Speakers and Accessories.** Full particulars and descriptive printed matter can be obtained from G.E.C. Stand, Radio Section, Gallery, Main Hall, or from The General Electric Co. Ltd., Magnet House, Kingsway, London, W.C.2, also from any Wireless Dealer.

NOTE.—GECoPHONE Wireless Products purchased to the value of £5 or over can be obtained on **HIRE PURCHASE** Terms. Ask for particulars.

Osram Valves

for

ECONOMICAL WIRELESS

Types for

BATTERY OPERATED AND ALL-ELECTRIC RECEIVERS.

STAND No. 465, RADIO SECTION, GALLERY, MAIN HALL

EPILOGUE

Exactly a century ago John Reith alerted readers of his autobiography to the dangers of being left behind. He acknowledged the suspicion with which some more educated people continued to view the new medium, but warned that 'the squire may suffer some embarrassment when he finds that his ploughman is better informed than he is on events of national significance'. It's an astonishingly dated passage, summoning up traces in 1920s' Britain of an almost feudal way of life, but Reith was, of course, correct in his prediction of how quickly the wireless set would come to be taken for granted as an essential and almost ubiquitous presence in the home: less than two decades.

Almost from the very start of audio broadcasting, there were some who predicted the arrival of that larger box of tricks, which was to usurp the place of radio by the fireside. By the end of 1958 the number of households with television exceeded the number of households that had only radio, rising to over 10 million as the decade came to a close and exploding beyond. Of course, the iconic image promoted by the early manufacturers and broadcasters of families gathered together to listen or watch by the fireside is no longer the norm, now that we consume so much of our broadcast media on the go or on our own. It was in the 1950s too that the arrival of the portable transistor radio

A magnificent range of wireless sets were offered by this retailer in 1935.

began this trend, particularly among younger listeners. A pocket-sized wireless gadget would certainly have impressed Guglielmo Marconi when he arrived at the London patent office in 1896. Listening on a smartphone, even more so. In fact, two decades are roughly what it has taken in the early twenty-first century for further startling transformations, not only of the technology but of the content of our everyday listening.

There are new methods for each individual listener to catch up on missed radio programmes or create their own personalized listening schedule, and new terms to describe these methods: 'listen again', 'listen before', 'audio on demand', 'audio on request', 'catch-up', 'downloads'. The advent of podcasting has been even more revolutionary – those now-no-longer-quite-such-new-kids-on-the-block sharing their enthusiasm for an endless array of mainstream or niche interests; their output with its ever-accessible and bingeable box-set appeal; their freedom from the constraints of formal radio; their altogether groundbreaking potential in allowing pretty much anyone with a laptop and a halfway-decent microphone to launch their voices into the ether.

I was almost thirty years into a wonderful career as a BBC radio producer when I began to understand how radically the medium in which I worked was changing. A small, personal epiphany helped spark the idea for this book and led me to explore more deeply what radio had been for my parents and grandparents, and just how it was changing for my generation and my daughter's generation. It happened in, I think, 2016, when I was attending a compulsory BBC course with a not-so-alluring subtitle, 'Using Meta-data When Writing Billings'.

Every time I produce a programme, I write a billing for it. This is the brief sales pitch which tells listeners what they might hear, whether generically for a live broadcast or in detail for a pre-recorded or built documentary (my speciality). These are printed in the *Radio Times* and all the other listings pages in local and national newspapers and magazines, as well as online of course. This is the traditional way of informing listeners about forthcoming programmes, encouraging them to tune in, and it dates back almost to the beginning of radio. But on this course we learned how growing numbers of listeners in general – and younger ones in particular – increasingly access their choices. These changes, we were told, were why it was essential to use keywords in our billings, as these would feed into the algorithms which today influence many would-be listeners. Listeners in general, we were told, were less and less likely just to switch on and listen out of habit or because they were inclined towards what Sir John Reith had dismissively termed 'tap listening' – that is to say, leaving the radio on in the background simply for companionship or as a sort of friendly noise. Nor were they likely to use a listings magazine or decide in advance to make a date with a programme. The term that the course leader used was 'mood navigation', by which she meant a listener's decision to follow their own individual identity, interests or feelings at that particular moment and search accordingly for what they want to hear. It is a massive shift from the era when my grandmother and her teenage daughters, one of them my mother, religiously tuned in every evening for the nine o'clock *News* during the Second World War; or when my father carefully drew rings with a fountain pen around certain concerts in the *Radio Times* and tried to make sure he was at home at those times to hear them. Far more recently, as anyone over a certain age will recall, radio still had an almost entirely fugitive quality and there was nothing uncommon about conversation along the lines of 'Did you hear so-and-so on the radio?' 'No, such a shame, I missed it...'

The course was intended to help us ensure that listeners found their way to our programmes and stayed listening to the networks on which they were broadcast rather than switching over or opting for different media entirely. Naturally I could see the sense in that, but I also immediately loved the term *mood navigation*. It seemed to capture a thrilling new sense of setting sail on the radio waves – audio waves, really – with nothing but one's individual feelings for a compass (though I was pretty sure I knew where Sir John would have stood on the matter...).

I love a podcast and I also appreciate the way that 'audio on demand' is giving a far longer shelf life to my programmes. Why, after all, produce a programme or series unless you want to give as many listeners as possible opportunities to hear it? There is anyway a good deal of crossover these days between linear radio and podcasts (though there are entire academic studies on their differences too). The catch-up facility is a huge bonus; plenty of one-off programmes and series originally made for linear radio can be downloaded; some radio programmes become podcast series and some podcast series are broadcast as radio programmes. But I myself will always be a fan of linear radio, broadcast live. For me, nothing ultimately replaces its reliability, its reassuring voice, the sense it brings of being part of an imagined – or, we might now say, virtual – community.

Perhaps rather quaintly, I have a particular affection for 'continuity': that is to say, the live linking material (as opposed to automated, pre-recorded content) which, on certain networks, is spoken by a range of announcers whose voices are so closely associated with that network that they sometimes remain in our inner ears long after they themselves are gone. Some programmes within the schedule are, of course, live, and some are pre-recorded, but nothing can replace the sense that someone is speaking right now, right there as I am right now, right here listening. For me, continuity is like the most intricate and expertly crafted but constantly changing setting for the individual gems which make up the programmes of the schedule.

Of course, the days of BBC Radio almost as a monopoly are long gone, and the audience is far more scattered than it was in the 1920s and 1930s, but all this contemporary change sheds a particularly timely light on those two decades in the not-too-distant past when broadcasting developed with such astonishing rapidity. This was when wireless became the go-to medium for most of the population, united in what they were hearing and in the knowledge of how many others were hearing it at the same time across the country. They were not following their moods or inclinations but simply switching on (or off). In our fragmented times – politically, culturally and socially segregated as we have become – these kinds of united experience are rare indeed, but deeply satisfying when they do occur: an international sports fixture, a glamorous royal occasion, the egregious downfall of a politician, the concluding episode of a compelling drama, the immersive coverage of a music festival... They are not to be dismissed lightly. While celebrating individuality, diversity and equality (farewell to Reith's division between squire and ploughman), there is today a deep need to seek out and share opportunities not only for listening but for listening together.

FINIS

Popular Wireless

TOPICAL NEWS AND NOTES

Wireless in Ireland.

"WIRELESS telephone licences are not to be issued in Northern Ireland until conditions are more settled," said Mr. Kellaway, the Postmaster-General, in a written Parliamentary answer to Major O'Neill.

In any case, if permission is eventually given, Ulster will be well within range of the Glasgow broadcasting station.

* * *

Journalistic Enterprise.

WIRELESS telephonic journalism was started in Holland as far back as last February, and so far the results have been very good. Fifty different newspaper subscribers of the Vasdiar Agency at Amsterdam, equipped with receiving sets, receive news throughout the day.

This is a good commencement, and although this service has not been welcomed over here, there is no denying the fact that it will eventually come—and come to stay.

* * *

The "Boom" in Canada.

WIRELESS is booming in Canada, as in the States. Many hundreds of receiving sets are in regular use in such places as Montreal, Toronto, and Winnipeg, by enthusiastic amateurs. Business concerns, specially in lumber operation, have been working over wide stretches of forest, using the wireless telephone with excellent results.

In the reporting of forest fires the radiophone has proved invaluable, and lumber companies are installing powerful apparatus connecting their offices with portable sets placed in the woods, thus opening up a wide field of commercial usefulness in the vast forests of Canada.

* * *

Recording Signals.

MESSAGES received by your wireless set when you are not present need not be lost. Signals can now be recorded on a special form of tape machine, or can be made to reproduce themselves on a gramophone record.

* * *

Broadcasting.

AN agreement has been arrived at between the Radio Communication Co. and the Metropolitan-Vickers Electrical Co. whereby these concerns propose to establish and jointly operate broadcasting stations.

Big things are expected of these two firms.

Schoolboy's Enterprise.

THE record for erecting the first amateur wireless station in North Devon has fallen to a schoolboy at the Devon County School, West Buckland, Master John B. Joyce, son of the Rev. Walter W. Joyce, Rector of Charles. The rector is the holder of the licence, and Joyce junior is the operator, the latter having been largely responsible for the construction of the apparatus, which consists of a single valve set, with necessary running coils, etc. The aerial is a single wire "inverted L," 80 feet long, height 35 feet, with 25 feet lead-in. The most interesting receptions are telephony and time signals from the Eiffel Tower. Both the 1,800 and the 2,600 metre transmissions of music and speech are plainly audible. The operator would be pleased to communicate at any time with any other genuine wireless amateur in the district.

* * *

Come, Birdy, Come!

"WHAT effect will the establishment of several new broadcasting stations have upon the birds?" asked one of our contemporaries a few days ago. "It is strange," thought the writer, "that an unseen influence, manipulated by man, can deflect the sure, instinctive flight of the birds. Nevertheless, it is a fact that our feathered friends are disturbed in a singular way by the wireless waves.

"Gulls appear to be the principal sufferers but large numbers of doves are in some way prevented from finding their way home when there are wireless stations in the line of flight. This strange phenomenon is attributed to some effect of the ether waves not yet understood."

I suggest the doves take out a licence at the nearest post-office and erect a direction finder at the earliest possible moment.

* * *

Amateur Wave Lengths.

THE decision of the committee which the Postmaster-General appointed has now removed some of the restrictions on the operations of wireless amateurs. A new wave length of 440 metres has been sanctioned for transmission, and the wireless amateur is to be exempt from inspection of his receiving station, and will no longer be restricted as to the length of receiving aerials.

* * *

Wireless on Trains.

THE Chicago, Milwaukee, and St. Paul Railway has equipped its trains with a radio system for the benefit of passengers. Arrangements have been made for the installation of complete radio systems in the club cars of the Pioneer Limited trains between Chicago, St. Paul, and Minneapolis. All news of the day will be received.

* * *

Wireless and Fishing Boats.

QUITE a number of North Sea trawlers and steam drifters are fitted with wireless telegraphy. Although their transmitting radius is small, the receiving capabilities are good. The purpose of the installations is said to be receiving, and most of the messages handled by them are in relation to the state of the fish market. When the market is glutted and the price of fish is low they are instructed to stay out and continue their fishing for another day or two. Similarly, when conditions force dealers to sell fish as manure, the fishing hauls can be diverted to other parts more fortunately situated.

* * *

Our Australian Confreres.

IN Australia to-day there are between 1,500 and 2,500 wireless experimenters, and it is anticipated that the number will grow very rapidly. It is

WHY WIRELESS IS POPULAR.

A fair amateur "tunes in" the Marconi concert.

ACKNOWLEDGEMENTS

Radio began for me in the home, so my first thanks are to my late parents, Enid and Frank Rubens, and to my five siblings, with whom I listened in childhood and have shared much over the years: Amanda, Giles, Kate, Quentin and Tilly. Above all others, I want to thank Giles, not only for sharing his thoughts, comments and suggestions but also for generously conducting a massive amount of research, particularly in the British Newspaper Archive. Without him, this book would be incomparably the poorer, and would contain a number of fanciful notions which he quite rightly urged me to rethink.

Emma Smith encouraged me from the start, and, on one of our many walks in the University Parks, helped redirect the project at a key point when I was flagging. I have gained so much from my radio collaborations with her and from our friendship. When I wrote to Bodley's Librarian, Richard Ovenden, proposing a possible book and exhibition on the subject of early radio in the home, I was extraordinarily fortunate that he took an immediate interest. He put me in touch with head of Exhibitions and Public Engagement Madeline Slaven, and with head of Bodleian Library Publishing, Samuel Fanous. How lucky I have been to work with both of them, with editor Janet Phillips, picture editor Leanda Shrimpton and also

This 'fair amateur' tunes in to a broadcast at a time when wireless was still a novelty and women were often discouraged from handling the set.

Suzy Gooch, Susie Foster and Dot Little. Working with Exhibitions Manager Sallyanne Gilchrist has been a visual education and a joy, while Ellen Brown's ability to turn up hidden Bodleian treasures has been awe-inspiring. For turning my scrappy manuscript (with its far from standard version of punctuation, honed over many years of writing for radio) into such a handsome book, I would like to thank Lucy Morton and Robin Gable.

My first real research was done in the remarkable John Johnson Collection of Printed Ephemera, which is held in the Bodleian Library. Its former librarian, Julie Anne Lambert, was outstandingly helpful from the start, and dug out some unexpected new material as she began to understand the dimensions of my project. This often meant that, while my fellow researchers were poring over exquisite volumes and priceless manuscripts, I was sifting through cigarette cards, theatre programmes and advertising brochures. Chris Fletcher, Keeper of Special Collections and Librarian in Charge of the Weston Library, shared his thoughts on the Winifred Gill Archive and also hosted the Friday coffee mornings in the Visiting Scholars Centre, which enabled me to mix with other, real, scholars and learn from them.

I am grateful to the very helpful staff at the BBC Written Archives in Caversham for checking out what it does and does not hold on Winifred Gill, thus helping me assess the importance of my Bodleian discovery. In Bristol, Garry Atterton of the Barton Hill History Group shared his enthusiasm for the district in which Gill and Hilda Jennings worked.

I feel hugely fortunate to have been awarded a Byrne–Bussey Marconi Visiting Fellowship in the History of Science, Technology and Communication at the Bodleian Library during the autumn of 2023. I am particularly grateful to Alexandra Franklin, who manages the Visiting Scholars Programme at the Bodleian, for her constant support and for the perceptive questions she asked; also to Rachel Naismith for her kindness and for sharing some family stories of radio listening. The librarians in the Weston Library Reading Room were unfailingly helpful and courteous, and I love working there.

At the History of Science Museum in the University of Oxford, Head of Research, Teaching and Collections, JC Niala, shared her enthusiasm for radio in general and for my project in particular. She also put me in touch with Ken Taylor, who, among much else, let me 'listen in' on headphones to an authentic 1920s-style signal and better imagine the early wireless experience.

At the Borthwick Institute for Archives, University of York, archivists Sally-Anne Shearn and Nicholas Melia provided much help with the Seebohm Rowntree material, including taking the trouble to digitize and share some precious early audio. I am grateful to my neighbour and friend Emily Baldwin for the introduction, as well as for enthusiastic conversations about the fiction of the 1920s and 1930s, and for leaving interesting books on my doorstep. Catherine Clarke at Felicity Bryan Associates took an early interest in the project, as did Rebecca Carter. Peter Mandler kindly advised me on how to go about researching the reality of early radio hobbyism. Seán Street not only read the chapter about the commercial opposition to the BBC but was also a kind guide in the early part of the whole undertaking, both as a scholar and as a poet and anthologist of radio poetry. Garry Atterton of the excellent Barton Hill History Group provided me with help about the area of Bristol where Winifred Gill and Hilda Jennings worked and conducted their research. When I wrote to her out of the blue, Siân Nicholas from the University of Aberystwyth very kindly helped me gauge the importance of the Winifred Gill material, while also introducing me to the significantly telling story about the early BBC executive who grandly declared that 'Nobody dines before eight!' Em Pritchard saved my eyesight by scouring copies of the *Radio Times* in various key years of the 1920s for cartoons and for the 1929 coverage of women's listening, drawing up a useful index of this fabulous resource. My friend Tony Morris encouraged me hugely with his constant enthusiasm and enriching conversations over the years. I would never have known about the wonderful writing of Maria Stepanova had it not been for my radio collaborations and friendship with her translator, the poet Sasha Dugdale.

During my thirty-five years at the BBC, many, many friends, colleagues and collaborators have, often unknowingly, contributed to this book. Paul Kobrak (a fellow member of Old Docs/New Tricks), took a constant interest and offered suggestions. Rob Ketteridge acted as a generous referee for my visiting fellowship. In my first BBC post, Susan Marling, now of Just Radio, taught me the golden rule that 'there are really only three questions: what, when and just how serious?' (though I am not entirely sure she is right about this...). Gillian Darley guided me towards some key books on interwar housing, and modelled for me what it means to be a truly tenacious researcher. I am grateful to Anne Smith, the Docs Gals, and to Giles Aspen, Jackie Margerum, Bob Nettles and the many studio managers who have helped me understand the technical side of radio over the years, as well as whipping my programmes into shape. I consulted Mark Damazer about the story of the Droitwich signal, while his successor as Controller of Radio Four, Mohit Bakaya, set me thinking about contemporary challenges to the future of linear radio. I have a special debt of gratitude to the incomparable Susan Rae, who, long ago, kindly allowed me to accompany her as she opened up Radio Four for the day, and shared the importance of wearing lipstick while broadcasting, even in an entirely aural medium. Working with James Naughtie on two big series has been a highlight of my career and I feel immensely privileged that he agreed to contribute the Foreword.

Many friends have helped and encouraged me along the way. With Cathy O'Neill and Melissa Marsh (3 like minds), Beck Fleming and Jules Wilkinson (The Stoop Group), and Rob and Fiona Lake, I have shared so much over the years, and I am deeply grateful to them all. For conversations about broadcasting and listening, which long ago helped sow the seeds of this project, I want to thank Rory Cellan-Jones, Diane Coyle, Pola Grzybowska and the late David Thomas. I first learned about making radio with Penny Boreham Saban: I still own a cassette copy of the magnificent (or so we thought) documentary which we produced together at the start of our long friendship. Meg Rosoff has been a constant support and wonderful friend.

For personal family memories related to early radio, I have been lucky to consult Adrian Davis, Rosemary Wolfson and Barbara and Robbie Barnett. In making a name for herself on BBC Radio Three as 'Jan from Cornwall', my sister-in-law Jan Wilson helped me think about individual interactions between listeners and networks. My brother-in-law Mike McCabe gave me historical insight into the period and (he has a reputation for this) carefully chose and sent me useful books. I would never have met Peter Lobbenberg or become friends with him had he not so generously helped with some early research, and I have Arthur Oppenheimer to thank for the introduction. I feel very fortunate to have the friendship and support of Kitty Hauser, Harriet Morley Hull and Claire Batten, as well as their exceptional visual expertise.

I hope that this book is not too full of errors, but such as there are belong to me and me alone.

I have dedicated it with love to Oliver Taplin and am grateful to him for introducing me to the idea of reception studies, for enthusiastically accompanying me on the journey, and for his utter steadfastness. It is also for Charis, to whom I owe so much, including an understanding of twenty-first-century changes in radio, audio and the whole experience of listening.

NOTES

EPIGRAPH

1. Mr Murgatroyd and Mr Winterbottom, BBC sketch, n.d., quoted in Barry Took, *Laughter in the Air: An Informal History of British Radio Comedy*, London: Robson Books, 1998, pp. 31–2.

FOREWORD

1. I am grateful to my sister Kate for reminding me of this piece of family history.
2. All statistics about radio take-up in the 1920s and 1930s are from Mark Pegg, *Broadcasting and Society, 1918–1939*, Beckenham: Croom Helm, 1983. The key table is on p. 7.
3. All editions of the *Radio Times* for the period of this book can be found at https://genome.ch.bbc.co.uk/issues.
4. Hilda Jennings and Winifred Gill, *Broadcasting in Everyday Life: A Social Survey of the Coming of Broadcasting*, London: British Broadcasting Corporation, 1939.
5. Kate Murphy, '"Careers for Women": BBC Women's Radio Programmes and the "Professional", 1923–1955', https://doi.org/10.1080/09612025.2022.2138018.

WHAT IS HOME WITHOUT A RADIO?

1. Arthur Mee, 'The Pleasure Telephone', *Strand Magazine*, 1898; Oxford, Bodleian Library, John Johnson Collection, PH148.
2. E.M. Delafield, *The Times*, Home supplement, 1935; reprinted in the introduction to *Ladies and Gentlemen in Victorian Fiction*, London: Hogarth Press, 1937, p. 12.
3. Asa Briggs, *The History of Broadcasting in the United Kingdom*, Volume 1: *The Birth of Broadcasting 1896–1927*, Oxford: Oxford University Press, 1961, p. 220.
4. Winifred Gill, red notebook, Oxford, Bodleian Library, Special Collections, MS. 6241/50, p. 51.
5. Briggs, *The History of Broadcasting*, vol. 1, p. 247.
6. Gill, red notebook, p. 113.
7. Alasdair Pinkerton, *Radio*, London: Reaktion Books, 2019, p. 53.

8. George Eliot, *Daniel Deronda*, London and Edinburgh: William Blackwood & Sons, 1889, opening of ch. 18, p. 145.
9. *Liverpool Echo*, 5 March 1920.
10. Maria Stepanova, *In Memory of Memory*, trans. Sasha Dugdale, London: Fitzcarraldo, 2021, p. 292; stress in original.
11. From 'The Victory', a poem written in tribute to Samuel Morse, 1872, quoted in Tom Standage, *The Victorian Internet: The Remarkable Story of the Telegraph and the Nineteenth Century's Online Pioneers*, London: Weidenfeld & Nicolson, 1998, p. 23.
12. Telephone flyer: https://collection.sciencemuseumgroup.org.uk/objects/co8088712/box-from-bell-telephone-and-terminal-panel-1877-box-telephone; accessed 29 July 2023.
13. John Johnson Collection, PH148.
14. Mee, 'The Pleasure Telephone'.
15. Edward Bellamy, *Looking Backwards 2000–1887*, Boston MA: Tickner, 1888.
16. Marcel Proust, *Correspondance*, ed. Philip Kolb, vols 10, 11 and 12, Paris: Plon, 1983–4; trans. by the author.
17. Tony Harrison, in conversation with the author.
18. *Daily Mirror*, 21 April 1922.
19. Delafield in *Ladies and Gentlemen in Victorian Fiction*, p. 12.
20. Michael Remson, *Septimus Winner: Two Lives in Music*, Lanham MD: Scarecrow Press, 2002, p. xvii, and the entire argument of the book.

READY OR NOT?

1. https://genome.ch.bbc.co.uk/page/c7c485105a7b4d0bb093b83cdb29beda?page=15.
2. Marconi Archives, History of Science Museum, University of Oxford, Inv.74444. Treviso version of Tarot cards (dating at least to the early 1880s and possibly far earlier), translated by Martin McLaughlin, Agnelli-Serena Professor of Italian Studies, Magdalen College, Oxford.
3. Cartoon by R. MacBeaney, 1912, reprinted in Seán Street, *A Concise History of British Radio 1922–2002*, Tiverton: Kelly Publications, 2002.
4. Cartoon by Leonard Raven-Hill, *Punch*, https://magazine.punch.co.uk/image/I0000TWpfoLPBGHM.
5. John Clarricoats, G6CL, *World at Their Fingertips: The Story of Amateur Radio in the United Kingdom and a History of the Radio Society of Great Britain*, Bedford: Radio Society of Great Britain, 1967, p. 10.
6. Ibid., p. 13.
7. P.P. Eckersley, *The Power Behind the Microphone*, London: Jonathan Cape, 1941, p. 27.
8. Clarricoats, *World at Their Fingertips*, p. 12.
9. Ibid., p. 32.
10. *The Marconigraph*, Oxford, Bodleian Library, John Johnson Collection, Wireless 3. (From April 1913 published as *The Wireless World*.)
11. Ibid.
12. *The Wireless World*, May 1913–March 1914, www.worldradiohistory.com/Wireless_World_Magazine.htm.
13. David Hendy, 'The Dreadful World of Edwardian Wireless', in Siân Nicholas and

Tom O'Malley, eds, *Moral Panics, Social Fears and the Media: Historical Perspectives*, Abingdon: Routledge, 2013, pp. 76–89.
14. Asa Briggs, *The History of Broadcasting in the United Kingdom*, Volume 1: *The Birth of Broadcasting 1896–1927*, Oxford: Oxford University Press, 1961, p. 41.
15. Ibid., p. 42.
16. *Liverpool Echo*, 20 December 1924.
17. www.bbc.co.uk/news/uk-england-essex.
18. *Liverpool Echo*, 20 December 1924.
19. R.T.B. Wynne, quoted in Briggs, *The History of Broadcasting*, p. 65.
20. P.P. Eckersley, *The Power Behind the Microphone*, London: Jonathan Cape, 1941, p. 43.
21. Oxford, Bodleian Library, Marconi Archives, MS. Marconi 312.
22. *The Sketch*, 20 May 1922, 5 March 1923, 29 August 1923.
23. Harry Pease, 'Listen in Virginia', quoted in Briggs, *The History of Broadcasting*, p. 1.
24. https://genome.ch.bbc.co.uk/page/c7c485105a7b4d0bb093b83cdb29beda?page=15.
25. *The Sketch*, 11 October 1922.
26. *The Sketch*, 17 October 1923.
27. *The Sketch*, 29 March 1922.
28. *The Tatler*, 19 September 1923.
29. *The Sketch*, 6 March 1923.
30. *Bayswater Chronicle*, 13 December 1924.
31. John Johnson Collection, Wireless 11.

WHAT ARE THE WILD WAVES SAYING?

1. Shaun Moores, '"The Box on the Dresser": Memories of Early Radio and Everyday Life', in Andrew Crisell, ed., *Radio: Critical Concepts in Media and Cultural Studies*, vol. 3, London: Routledge, 2008, p. 28.
2. *Daily Mail*, 24 April 1924.
3. Oxford, Bodleian Library, Marconi Archives, MS. Marconi 312.
4. *Daily Mail*, 24 April 1924.
5. Ibid.
6. *The Sphere*, 3 May 1924.
7. Moores, '"The Box on the Dresser"', p. 28.
8. Ibid., p. 2.
9. *The Sphere*, 29 March 1924.
10. *The Sketch*, 14 March 1923.
11. Mark Girouard, *The Victorian Country House*, New Haven CT and London: Yale University Press, 1979, p. 307.
12. Moores, '"The Box on the Dresser"', p. 29.
13. https://genome.ch.bbc.co.uk/page/f131b3fca2254dab8ff4b8e71ac15d3a?page=17.
14. *Manchester Evening Chronicle*, March 1924, cited in Paddy Scannell and David Cardiff, *A Social History of British Broadcasting*, Volume 1: *1922–1939, Serving the Nation*, Oxford: Basil Blackwell, 1991, p. 313.
15. *Punch* cartoon, cited in Scannell and Cardiff, *A Social History*, p. 331.

16. Scannell and Cardiff, *A Social History*, p. 358.
17. Asa Briggs, *The History of Broadcasting in the United Kingdom*, Volume 1: *The Birth of Broadcasting 1896–1927*, Oxford: Oxford University Press, 1961, p. 234.
18. Lesley Bailey, BBC programme *Scrapbook for 1922*, cited in Briggs, *The History of Broadcasting*, p. 128.
19. Walter S. Masterman, *2LO*, London: Victor Gollancz,1928.
20. https://genome.ch.bbc.co.uk/page/34698110 3a944dd48925fc685d846c87?page=4.
21. https://genome.ch.bbc.co.uk/page/be558ab645fd4cf6b1d332ed99174387?page=23.
22. https://genome.ch.bbc.co.uk/page/c7c485105a7b4d0bb093b83cdb29beda?page=13.
23. J.C.W. Reith, *Broadcast Over Britain*, London: Hodder & Stoughton, 1924, p. 77.
24. Ibid.
25. *World Radio*, 17 December 1937, p. 33; Oxford, Bodleian Library, John Johnson Collection, Wireless 11.
26. C.A. Lewis, *Broadcasting from Within*, London: George Newnes, p. 118.
27. Moores, '"The Box on the Dresser"', pp. 29–30.
28. Reith, *Broadcast Over Britain*, p. 223.
29. https://genome.ch.bbc.co.uk/page/32a960e9fd374d2abdbad56ceaef7791.
30. Susan Briggs, *Those Radio Times*, London: Weidenfeld & Nicolson, 1981, p. 35.
31. Lesley Bailey, BBC programme *Scrapbook for 1925*, cited in Briggs, *The History of Broadcasting*, p. 13.

DAVENTRY CALLING

1. Winifred Gill, red notebook, Oxford, Bodleian Library, Special Collections, MS .6241/50, p. 51.
2. https://genome.ch.bbc.co.uk/page/2bb6970a276543158c48577a1b42ef10?page=25.
3. Alfred Noyes, *The Dane Tree*, The Society of Authors (Representatives of the Literary Estate of Alfred Noyes), reprinted in Seán Street, *Radio Waves: Poems Celebrating the Wireless*, London: Enitharmon Press, 2004.
4. Hilda Jennings and Winifred Gill, *Broadcasting in Everyday Life: A Social Survey of the Coming of Broacasting*, London: British Broadcasting Corporation, 1939, pp. 9–11.
5. *First Chairman of Convocation: Biographies of the Former Chairmen of Convocation*, Bristol University Alumni, archived from the original (both pdf) on 1 May 2014; retrieved 14 March 2013.
6. Gill, red notebook, p. 4.
7. Ibid., p. 118.
8. Ibid., p. 119.
9. Ibid., p. 10.
10. Ibid., p. 96.
11. *Motherwell Times*, 15 June 1928.
12. Gill, red notebook, p. 34.
13. BBC memo, 1924, Oxford, Bodleian Library, Marconi Archives, MS. Marconi 313.

14. https://genome.ch.bbc.co.uk/page/15ef1f23bd864156a73f1c1762e96de4?page=3.
15. *The Tatler*, 6 April 1932.
16. www.bbc.co.uk/blogs/radio4/2012/09/and_now_an_urgent_sos_message.html?fbclid=IwAR1xcXOixtowgzgq3DJard3smBYUBOgIP2ZmppPGL5yZkXMqA28F-ROgbic.
17. https://genome.ch.bbc.co.uk/page/c314e097c80a4c3ab8cf50776775a6ee?page=22.
18. *Hull Daily Mail*, 27 December 1937.
19. Carol Anne Duffy, 'Prayer', in *Mean Time*, London: Anvil Press Poetry, 1993.
20. https://connectedhistoriesofthebbc.org/play/?id=49.
21. Oxford, Bodleian Library, John Johnson Collection, Wireless 11.
22. https://genome.ch.bbc.co.uk/page/496933c47da448ee9ed6e5507994c36f?page=3.
23. www.bbc.co.uk/blogs/bbchistoryresearch/entries/61c1c1bb-0d83-45bc-be16-8179c8f4d6b5.
24. J.C.W. Reith, *Broadcast Over Britain*, London: Hodder & Stoughton, 1924, p. 185.
25. https://eprints.lincoln.ac.uk/id/eprint/52291/7/Healy%20Zara%20-%20Media%20-%20%20June%202022%20.pdf.
26. C.A. Lewis, quoted in Susan Briggs, *Those Radio Times*, London: Weidenfeld & Nicolson, 1981, p. 93.

BUILDING JERUSALEM

1. Winifred Gill, red notebook, Oxford, Bodleian Library, Special Collections, MS. 6241/50, p. 5.
2. David Hendy, *The BBC: A People's History*, London: Profile Books, 2022, p. 109.
3. Ibid., p. 110.
4. Asa Briggs, *The History of Broadcasting in the United Kingdom*, Volume 1: *The Birth of Broadcasting 1896–1927*, Oxford: Oxford University Press, 1961, p. 338.
5. Rachelle Hope Saltzman, *A Lark for the Sake of Their Country: The 1926 General Strike Volunteers in Folklore and Memory*, Manchester: Manchester University Press, 2012, p. 109.
6. Beatrice Webb, *Diaries, 1924–31*, ed. M. Cole, London: Longman, 1956, cited in Mark Pegg, *Broadcasting and Society, 1918–1939*, Beckenham: Croom Helm, 1983.
7. Saltzman, *A Lark for the Sake of Their Country*, p. 87.
8. Ibid., p. 45.
9. Briggs, *The History of Broadcasting*, p. 338.
10. Explored (without final confirmation) by Peter Hennessy in an archive item broadcast on the *Today* programme, BBC Radio Four, 11 December 2023.
11. Robert Lynd in the *New Statesman*, 8 May 1926, p. 103, cited in Pegg, *Broadcasting and Society*, p. 108.
12. Julian Symons, *The General Strike*, London: Cresset Press, p. 181.
13. Saltzman, *A Lark for the Sake of Their Country*, p. 45; Hendy, *The BBC*, p. 111.
14. https://genome.ch.bbc.co.uk/page/f3a3c5b748cb40779c971ee883033cbf?page=3.
15. *Birmingham Evening Despatch*, July 1924 (day not legible).

16. https://genome.ch.bbc.co.uk/page/c5c24f659afc45b190d862eea0573227?page=2.
17. Briggs, *The History of Broadcasting*, pp. 266–7.
18. *Broadcast English I. Recommendations to Announcers Regarding Certain Words of Doubtful Pronunciation*, with an Introduction by A. Lloyd James, London: British Broadcasting Corporation, 1928, pp. 5–20, in Oxford, Bodleian Library, John Johnson Collection, Wireless 6.
19. *Broadcast English II. The BBC's Recommendations to Announcers Regarding the Pronunciation of Some English Place Names*, ed. A. Lloyd James, London: British Broadcasting Corporation, 1930, p. 7, in John Johnson Collection, Wireless 6.
20. Ibid.
21. Gill, red notebook, p. 40.
22. Ibid., p. 21.
23. Ibid., p. 108.
24. Ibid., p. 39.
25. *The Evidence Regarding Broadcast Speech Training*, London: British Broadcasting Corporation, n.d., p. 27, in John Johnson Collection, Wireless 6.
26. Ibid.
27. Hilda Jennings and Winifred Gill, *Broadcasting in Everyday Life: A Social Survey of the Coming of Broadcasting*, London: British Broadcasting Corporation, 1939, p. 19.
28. Gill, red notebook, p. 49.
29. www.youtube.com/watch?v=vsQYXAEpn1Y.
30. www.youtube.com/watch?v=w3qBGoehIo8&t=29s.
31. https://journals.openedition.org/rfcb/7681; and, for an illuminating obituary of Shapley, www.theguardian.com/news/1999/mar/15/guardianobituaries.
32. https://genome.ch.bbc.co.uk/page/768d74e8d22f4d1face49ae79dcff14c?page=10.
33. Gill, red notebook, p. 103.
34. www.youtube.com/watch?v=w3qBGoehIo8&t=29s.

THE 'FLAPPER ELECTION'

1. Letter to Arthur Burrows from R.H. White, 7 September 1922, in the Marconi Archives, cited in Asa Briggs, *The History of Broadcasting in the United Kingdom*, Volume 1: *The Birth of Broadcasting 1896–1927*, Oxford: Oxford University Press, 1961, p. 69.
2. R. Murray-Leslie, *The Times*, 20 February 1908, p. 15.
3. *The Broadcaster*, cited in Susan Briggs, *Those Radio Times*, London: Weidenfeld & Nicolson, 1981, p. 29.
4. Oxford, Bodleian Library, John Johnson Collection, Wireless 10.
5. *The Sketch*, 11 October 1922.
6. This paragraph draws heavily on Deborah Sugg Ryan, *Ideal Homes: Uncovering the History and Design of the Interwar House*, Manchester: Manchester University Press, 2018, pp. 58–9, 76.
7. https://genome.ch.bbc.co.uk/page/42fc8ac3b1314c889270dd883abab4c2?page=2.

8. Michael Remson, *Septimus Winner: Two Lives in Music*, Lanham MD: Scarecrow Press, 2002, the entire argument of the book.
9. https://genome.ch.bbc.co.uk/page/89ba974f033f437ead84c87cac6d534b?page=9.
10. Winifred Gill, red notebook, Oxford, Bodleian Library, Special Collections, MS. 6241/50, p. 73.
11. Ibid., p. 72.
12. Shaun Moores, '"The Box on the Dresser": Memories of Early Radio and Everyday Life', in Andrew Crisell, ed., *Radio: Critical Concepts in Media and Cultural Studies*, vol. 3, London: Routledge, 2008, p. 36.
13. Gill, red notebook, p. 27, and https://genome.ch.bbc.co.uk/page/89ba974f033f437ead84c87cac6d534b? page=9.
14. Gill, red notebook, p. 38.
15. https://genome.ch.bbc.co.uk/page/faf428728e9c425f87c90d29744f64c3?page=5.
16. Gill, red notebook, p. 13.
17. Ibid., p. 31.
18. https://genome.ch.bbc.co.uk/page/6ebf03a4967942438a1964a4637ca101?page=4.
19. Charlotte Higgins,*This New Noise: The Extraordinary Birth and Troubled Life of the BBC*, London: Guardian Books/Faber & Faber, 2015, p. 16.
20. Kate Murphy, *From Women's Hour to Other Women's Lives: BBC Talks for Women and the Women Who Made Them, 1923–1939*, https://eprints.bournemouth.ac.uk/33052, p. 3 (accessed 12 January 2024).
21. Details and extensive analysis of the fascinating life and work of Hilda Matheson are to be found in Michael Carney and Kate Murphy, *Hilda Matheson: A Life of Secrets and Broadcasts*, Bath: Handheld Press, 2023, particularly in the long concluding essay by Kate Murphy, 'Hilda Matheson: With and for Women, 1920–1931'. See also Kate Murphy, *From Women's Hour to Other Women's Lives: BBC Talks for Women and the Women Who Made Them, 1923–39*, Abingdon: Routledge, 2014.
22. Higgins, *This New Noise*, p. 28.
23. Carney and Murphy, *Hilda Matheson*, p. 82.
24. Gill, red notebook, p. 102.
25. Ibid.
26. Ibid., p. 24.
27. Maggie Andrews, *Domesticating the Airwaves: Broadcasting, Domesticity and Femininity*, London: Continuum, 2012, p. 40.
28. *Home, Health and Garden: BBC Household Talks 1928*, London: British Broadcasting Corporation, in Oxford, Bodleian Library, John Johnson Collection, Wireless 8.
29. Gill, red notebook, p. 112.
30. Ibid., p. 9.
31. Ibid., p. 2.
32. www.nypl.org/events/tours/audio-guides/virginia-woolf-modern-mind/item/13614.
33. E.M. Delafield, *The Provincial Lady Goes Further*, London: Macmillan, 1932, pp. 3–4.

34. https://genome.ch.bbc.co.uk/page/f0a2accf36b7450aa0086c14143658a1?page=25.
35. All statistics about the wireless trade in the 1920s and 1930s are from Mark Pegg, *Broadcasting and Society, 1918–1939*, Beckenham: Croom Helm, 1983, p. 52.
36. Gill, red notebook, p. 53.
37. Ibid., p. 20.
38. Ibid., p. 117.
39. Ibid., p. 61.
40. Moores, '"The Box on the Dresser"', p. 30.
41. Ibid., p. 30.
42. Ibid., p. 25.
43. Winifred Gill, green notebook, Oxford, Bodleian Library, Special Collections, MS. 6241/50, n. pag.
44. Gill, red notebook, p. 81.
45. Oxford, Bodleian Library, John Johnson Collection, Wireless 1.
46. Hilda Jennings and Winifred Gill, *Broadcasting in Everyday Life: A Social Survey of the Coming of Broadcasting*, London: British Broadcasting Corporation, 1939, p. 22.
47. Gill, red notebook, p. 55.
48. Murphy, *From Women's Hour to Other Women's Lives*, p. 14.
49. Gill, red notebook, p. 8.
50. Mass Observation Archive – DS81, 7 July 1937, quoted in Andrews, *Domesticating the Airwaves*, p. 26.
51. Carney and Murphy, *Hilda Matheson*, p. 225.

ONLY VOICES OUT OF THE AIR

1. Winifred Gill, red notebook, Oxford, Bodleian Library, Special Collections, MS. 6241/50, p. 1.
2. First Empire Address by King George V, at www.bbc.com/historyofthebbc/anniversaries/december/christmas-message#:~:text=In%20the%20speech%2C%20which%20was,my%20peoples%20throughout%20the%20Empire.%22.
3. Ibid.
4. *The Sketch*, 14 March 1927.
5. *The Tatler*, 16 October 1929.
6. *The Sphere*, 29 January 1927.
7. George Orwell, *The Lion and the Unicorn: The Collected Essays, Journalism and Letters of George Orwell*, Volume I: *An Age Like This*, London: Secker & Warburg, 1968, p. 77.
8. *Radio Times*, 20 January 1928.
9. *The Sketch*, 11 January 1930.
10. *Daily Mirror*, 6 June 1929.
11. Shaun Moores, '"The Box on the Dresser": Memories of Early Radio and Everyday Life', in Andrew Crisell, ed., *Radio: Critical Concepts in Media and Cultural Studies*, vol. 3, London: Routledge, 2008, p. 25.
12. Gill, red notebook, p. 9.
13. Ibid., p. 11.
14. Ibid., p. 41.

15. Ibid.
16. Ibid., p. 76.
17. *Hartlepool Northern Daily Mail*, 27 July 1928.
18. Gill, red notebook, p. 50.
19. Ibid., p. 5.
20. Ibid., p. 50.
21. Hilda Jennings and Winifred Gill, *Broadcasting in Everyday Life: A Social Survey of the Coming of Broadcasting*, London: British Broadcasting Corporation, 1939, p. 37.
22. Ibid., p. 38.
23. *Liverpool Echo*, 20 December 1924.
24. Ibid.
25. Asa Briggs, *The History of Broadcasting in the United Kingdom*, Volume 1: *The Birth of Broadcasting 1896–1927*, Oxford: Oxford University Press, 1961, p. 13.
26. *Liverpool Echo*, 20 December 1924.
27. Gill, red notebook, p. 12.
28. Ibid., p. 54.
29. Ibid., p. 119.
30. *The Sketch*, 11 January 1933.
31. *The Listener*, 19 January 1939, in Oxford, Bodleian Library, John Johnson Collection, Wireless 4.
32. Brian Friel, *Dancing at Lughnasa*, London: Faber & Faber, 1990, p. 3.
33. Ibid., p. 31.
34. Gill, red notebook, p. 69.
35. Ibid., p. 26.
36. Ibid., p. 35.
37. Thomas Hajkowski, *The BBC and National Identity in Britain, 1922–53*, Manchester: Manchester University Press, 2010, p. 21.
38. www.bbc.co.uk/worldservice/specials/1122_75_years/page2.shtml.
39. Penelope Fitzgerald, *Human Voices*, London: Collins, 1980; republished with a new introduction by Mark Damazer, London: Fourth Estate, 2014, p. 90.
40. *Architectural Review*, May 1932.
41. *Daily Mirror*, 13 May 1932.
42. Ibid.
43. John Johnson Collection, Wireless 9.
44. Details of Broadcasting House from *A Technical Description of Broadcasting House*, London: British Broadcasting Corporation, 1939, in Oxford, Bodleian Library, John Johnson Collection, Wireless.

WE ARE THE OVALTINEYS

1. Winifred Gill, red notebook, Oxford, Bodleian Library, Special Collections, MS. 6241/50, p. 51.
2. www.bbc.co.uk/blogs/bbchistoryresearch/entries/6c3501c1-cae7-4494-8276-13a92b10e69e.
3. *Radio Pictorial* editions can be accessed at www.worldradiohistory.com/Radio_Pictorial.htm.
4. Gill, red notebook, p. 53.

5. J.C.W. Reith, *Broadcast Over Britain*, London: Hodder & Stoughton, 1924, p. 193.
6. BBC Control Board Minutes, BBC Written Archives Centre, Caversham, 14 July and 16 November 1926, cited in Michael Carney and Kate Murphy, *Hilda Matheson, A Life of Secrets and Broadcasts*, Bath: Handheld Press, 2023, p. 3.
7. B. Seebohm Rowntree, *Poverty and Progress*, London: Longman, 1941, p. 469.
8. *Leisure Time Activities Report*, 1936, Rowntree Family Papers, Borthwick Institute for Archives, University of York, RFAM/BSR/UOY/13/21.
9. Winifred Gill, green notebook, Oxford, Bodleian Library, Special Collections, MS. 6241/50, n. pag.
10. John Mackintosh & Sons Ltd Archive, Borthwick Institute for Archives, University of York, M/S/ELF/62.
11. Susan Briggs, *Those Radio Times*, London: Weidenfeld & Nicolson, 1981, cited on p. 102.
12. Ibid., cited on p. 95.
13. Ibid.
14. E.M. Delafield, *Ladies and Gentlemen in Victorian Fiction*, London: Hogarth Press, 1937, p. 13.
15. *BBC Year Book*, London: British Broadcasting Corporation, 1930.
16. George Orwell, *The Road to Wigan Pier*, London: Victor Gollancz, 1937, p. 92.
17. *The Sphere*, 29 January 1927.
18. *The Sphere*, 8 February, 1930.
19. Mark Pegg, *Broadcasting and Society, 1918–1939*, Beckenham: Croom Helm, 1983, cited on p. 207.
20. Reith, *Broadcast Over Britain*, p. 34.
21. 'The Wireless Talk', *Huddersfield Daily Examiner*, 7 January 1930, including observation about listener choices in the home.
22. *New Ventures in Broadcasting*, London: British Broadcasting Corporation, 1928, in Oxford, Bodleian Library, John Johnson Collection, Wireless 6.
23. https://genome.ch.bbc.co.uk/page/301c449cc3f144a6a75307e0f82c49a7.
24. John Johnson Collection, Wireless 6, BBC booklet on *The Evidence Regarding Broadcast Speech Training*. A critical tone was increasingly common in popular newspaper wireless columns, and also in magazines such as *The Wireless League Gazette*, the first issue of which is in John Johnson Collection, Wireless 6.
25. 'What Radio Listeners Prefer', *Sunderland Daily Echo and Shipping Gazette*, 26 May 1933, cites the manufacturers' survey.
26. H.C. Colles, 'Walford Davies', *Music & Letters*, vol. 22, no. 3, July 1941, Oxford: Oxford University Press, p. 199.
27. Gill, red notebook, p. 111.
28. Ibid., p. 91.
29. Ibid., p. 31.
30. Pegg, *Broadcasting and Society*, p. 210.
31. *New Ventures in Broadcasting*.
32. David Hendy, *The BBC: A People's History*, London: Profile Books, 2022, p. 143, for details of the Norman Long debacle and the BBC's move towards 'Stardust'.
33. Paddy Scannell and David Cardiff, *A Social History of British Broadcasting*, Volume 1: *1922–1939, Serving the Nation*, Oxford: Basil Blackwell, 1991, p. 267, and the

entire chapter on 'Styles of Variety', pp. 246–73, gives a detailed account of the development of variety at the BBC.

34. Gill, red notebook, p. 68.
35. BBC Radio Four programme, *In Town Last Night*, broadcast 2003, CD recording in the personal archive of the author.
36. Cited in Barry Took, *Laughter in the Air: An Informal History of British Radio Comedy*, London: Robson Books, 1981, pp. 31–2.
37. I am grateful to Seán Street for this pithy phrase.

THERE'S A WOMAN IN HULL NOT SINGING

1. Humphrey Jennings and Charles Madge, eds, *Mass-Observation Day-Survey, May 12th 1937*, London: Faber & Faber, 1937, p. 275, CL 71.
2. *Nature*, vol. 137, no. 141, 1936; https://doi.org/10.1038/137141a0.
3. http://news.bbc.co.uk/1/hi/uk/2707489.stm.
4. Robert Hawes, *Radio Art*, London: Green Wood, 1991, p. 40.
5. Ibid., p. 36.
6. Shaun Moores, '"The Box on the Dresser"': Memories of Early Radio and Everyday Life', in Andrew Crisell, ed., *Radio: Critical Concepts in Media and Cultural Studies*, vol. 3, London: Routledge, 2008, p. 29.
7. Jennings and Madge, eds, *Mass-Observation*, p. 326.
8. Ibid., p. 272.
9. Ibid., p. 273.
10. Ibid., p. 279.
11. Ibid., p. 278.
12. Ibid., p. 293.
13. Ibid., p. 284.
14. Ibid., p. 280.
15. Ibid., p. 294.
16. Ibid., p. 275.
17. Ibid., p. 282.
18. Ibid., p. 281.
19. Graham Mytton, *Handbook on Radio and Television Audience Research*, https://unesdoc.unesco.org/ark:/48223/pf0000124231 p34.
20. Paddy Scannell and David Cardiff, *A Social History of British Broadcasting*, Volume 1: *1922–1939, Serving the Nation*, Oxford: Basil Blackwell, 1991, p. 273.
21. Winifred Gill, red notebook, Oxford, Bodleian Library, Special Collections, MS. 6241/50, p. 96.
22. Ibid., p. 12.
23. J.C.W. Reith, *Broadcast Over Britain*, London: Hodder & Stoughton, 1924, p. 34.

THE HOME SERVICE

1. Winifred Gill, red notebook, Oxford, Bodleian Library, Special Collections, MS. 6241/50, p. 25.
2. https://genome.ch.bbc.co.uk/page/5d09a0a3a5734739af0dbae2e1b9b8a0?page=2.
3. https://genome.ch.bbc.co.uk/page/c1d4bdd4279f45a7bedf9f9dd3c22919.
4. Gill, red notebook, p. 76.

5. Ibid., p. 106.
6. Ibid., p. 62.
7. Ibid., p. 35.
8. Ibid., p. 46.
9. Ibid., pp. 71–2.
10. Hilda Jennings and Winifred Gill, *Broadcasting in Everyday Life: A Social Survey of the Coming of Broadcasting*, London: British Broadcasting Corporation, 1939, p. 31.
11. Gill, red notebook, p. 57.
12. Winifred Gill, green notebook, Oxford, Bodleian Library, Special Collections, MS. 6241/50, n. pag.
13. Gill, red notebook, p. 112.
14. Ibid., p. 27.
15. *Punch*, 21 September 1938, reprinted in Paddy Scannell and David Cardiff, *A Social History of British Broadcasting*, Volume 1: *1922–1939, Serving the Nation*, Oxford: Basil Blackwell, 1991, p. 129.
16. Miss French, quoted in Dorothy Sheridan, ed., *Wartime Women: A Mass-Observation Anthology 1937 – 45*, Abergavenny: Phoenix, 2009, p. 29.
17. Virginia Woolf, *A Writer's Diary*, London: Hogarth Press, 1953, pp. 300–301.
18. Gill, red notebook, p. 14.
19. Ibid., p. 57.
20. Mark Pegg, *Broadcasting and Society, 1918–1939*, Beckenham: Croom Helm, 1983. p. 151.
21. Gill, red notebook, p. 100.
22. Ibid., p. 69.
23. Frank McDonough, *Neville Chamberlain, Appeasement, and the British Road to War*, Manchester: Manchester University Press, 1998, p. 126.
24. Gill, red notebook, p. 100.
25. Ibid., p. 89.
26. Ibid., p. 91.
27. Cited in Pegg, *Broadcasting and Society*, p. 153.
28. E.M. Delafield, *The Provincial Lady in War-Time*, London: Macmillan, 1940, p. 24.
29. Pegg, *Broadcasting and Society*, p. 126.
30. Penelope Fitzgerald, *Human Voices*, London: Collins, 1980; republished with a new introduction by Mark Damazer, London: Fourth Estate, 2014, p. 43.
31. Ibid., p. 16.
32. J. Simon Potter, *This is the BBC: Entertaining the Nation, Speaking for Britain? 1922–2022*, Oxford: Oxford University Press, 2022, pp. 77–84.
33. Fitzgerald, *Human Voices*, p. 103.
34. https://genome.ch.bbc.co.uk/page/ed159be73b134c399aab4591c6209e8b.

FURTHER READING

Andrews, Maggie, *Domesticating the Airwaves: Broadcasting, Domesticity and Femininity*, London: Continuum, 2012.

Artmonsky, Ruth, *Art for the Ear: Forty Years of Illustration for BBC Radio Publications*, London: Artmonsky Arts, 2015.

Beauman, Nicola, *A Very Great Profession: The Woman's Novel 1914–39*, London: Virago, 1983.

Briggs, Asa, *The History of Broadcasting in the United Kingdom*, Volume 1: *The Birth of Broadcasting 1896–1927*, Oxford: Oxford University Press, 1961.

Briggs, Susan, *Those Radio Times*, London: Weidenfeld & Nicolson, 1981.

Carney, Michael, and Kate Murphy, *Hilda Matheson, A Life of Secrets and Broadcasts*, Bath: Handheld Press, 2023.

Clarricoats, John, G6CL, *World at Their Fingertips: The Story of Amateur Radio in the United Kingdom and a History of the Radio Society of Great Britain*, Bedford: Radio Society of Great Britain, 1967.

Delafield, E.M., *Diary of a Provincial Lady*, London: Macmillan, 1930.

Delafield, E.M., *The Provincial Lady Goes Further*, London: Macmillan, 1932.

Delafield, E.M., *Ladies and Gentlemen in Victorian Fiction*, London: Hogarth Press, 1937.

Delafield, E.M., *The Provincial Lady in War-Time*, London: Macmillan, 1940.

Driver, David, *The Art of Radio Times*, London: BBC Publications, 1981.

Eckersley, P.P., *The Power Behind the Microphone*, London: Jonathan Cape, 1941.

Elmes, Simon, *And Now on Radio 4*, London: Penguin Random House, 2007.

Fitzgerald, Penelope, *Human Voices*, London: Collins, 1980; republished with a new introduction by Mark Damazer, London: Fourth Estate, 2014.

Friel, Brian, *Dancing at Lughnasa*, London: Faber & Faber, 1990.

Hajkowski, Thomas, *The BBC and National Identity in Britain, 1922–53*, Manchester: Manchester University Press, 2010.

Hawes, Robert, *Radio Art*, London: Green Wood, 1991.

Hendy, David, *Noise: A Human History of Sound and Listening*, London: Profile Books, 2013.

Hendy, David, *The BBC: A People's History*, London: Profile Books, 2022.

Higgins, Charlotte, *This New Noise: The Extraordinary Birth and Troubled Life of the BBC*, London: Guardian Books/Faber & Faber, 2015.
Hill, Jonathan, *Old Radio Sets*, Oxford: Shire, 1993.
Jennings, Hilda, and Winifred Gill, *Broadcasting in Everyday Life: A Social Survey of the Coming of Broadcasting*, London: British Broadcasting Corporation, 1939.
Jennings, Humphrey and Charles Madge, eds, *Mass-Observation Day-Survey, May 12th 1937*, London: Faber & Faber, 1937.
Lewis, C.A., *Broadcasting from Within*, London: George Newnes, 1924.
Masterman, Walter S., *2LO*, London: Victor Gollancz, 1928; reprinted Vancleave MS, Ramble House, 2020.
Moores, Shaun, '"The Box on the Dresser": Memories of Early Radio and Everyday Life', in Andrew Crisell, ed., *Radio: Critical Concepts in Media and Cultural Studies*, vol. 3, London: Routledge, 2008.
Murphy, Kate, *Behind the Wireless: A History of Early Women at the BBC*, London: Palgrave Macmillan, 2016.
Nicholas, Siân, and Tom O'Malley, eds., *Moral Panics, Social Fears and the Media: Historical Perspectives*, Abingdon: Routledge, 2013.
Orwell, George, *The Road to Wigan Pier*, London: Victor Gollancz, 1937.
Pegg, Mark, *Broadcasting and Society, 1918–1939*, Beckenham: Croom Helm, 1983.
Pinkerton, Alasdair, *Radio*, London: Reaktion Books, 2019.
Potter, J. Simon, *This is the BBC: Entertaining the Nation, Speaking for Britain? 1922–2022*, Oxford: Oxford University Press, 2022.
Reith, J.C.W., *Broadcast Over Britain*, London: Hodder & Stoughton, 1924.
Remson, Michael, *Septimus Winner: Two Lives in Music*, Lanham MD: Scarecrow Press, 2002.
Saltzman, Rachelle Hope, *A Lark for the Sake of Their Country: The 1926 General Strike Volunteers in Folklore and Memory*, Manchester: Manchester University Press, 2012.
Sayers, Dorothy L., *Busman's Honeymoon*, London: Victor Gollancz,1937.
Scannell, Paddy, and David Cardiff, *A Social History of British Broadcasting*, Volume 1: *1922–1939, Serving the Nation*, Oxford: Basil Blackwell, 1991.
Seatter, Robert, *Broadcasting Britain: 100 Years of the BBC*, London: Dorling Kindersley, 2022.
Silvey, Robert, *Who's Listening? The Story of BBC Audience Research*, London: George Allen and Unwin, 1974.
Standage, Tom, *The Victorian Internet: The Remarkable Story of the Telegraph and the Nineteenth Century's Online Pioneers*, London: Weidenfeld & Nicolson, 1998.
Stepanova, Maria, *In Memory of Memory*, trans. Sasha Dugdale, London: Fitzcarraldo, 2021.
Street, Seán, *A Concise History of British Radio 1922–2002*, Tiverton: 2002; reprinted (with corrections) 2004.
Street, Seán, *Radio Waves: Poems Celebrating the Wireless*, London: Enitharmon Press, 2004.
Sugg Ryan, Deborah, *Ideal Homes: Uncovering the History and Design of the Interwar House*, Manchester: Manchester University Press, 2018.
Wilkinson, Ellen, *Clash*, ed. Ian Haywood and Maroula Joannou, Nottingham: Trent Editions, 2004.
Woolf, Virginia, *A Writer's Diary*, London: Hogarth Press, 1953.

PICTURE CREDITS

viii *The Radio Beacon*, February 1926. Oxford, Bodleian Library, John Johnson Collection: Wireless box 5.
xii Advertisement for *Popular Wireless*, 1922. Oxford, Bodleian Library, John Johnson Collection: Prospectuses of Journals 42 (20).
4 Library of Congress, Prints and Photographs Collection, Photo by Underwood & Underwood.
6 Oxford, Bodleian Library, MSS. 6241/50.
12 *Radio Times*, 21 December 1923. Oxford, Bodleian Library, Per. 247933 c.10, vol. 1.
15 © National Portrait Gallery, London P1700 (55).
18 © History of Science Museum, University of Oxford, Inv. 28052.
23 Poster by Jules Chéret, 1896. New York Public Library, The Miriam and Ira D. Wallach Division of Art, Prints and Photographs. 118544.
24 Oxford, Bodleian Library, John Johnson Collection, Postal History 148.
29 Courtesy of BT Group Archives.
30 *Radio Times*, 5 September 1926. Oxford, Bodleian Library, Per. 247933 c.10, vol. 12, p. 410.
32 *The Marconigraph*, April 1911. Oxford, Bodleian Library, John Johnson Collection: Wireless box 3.
34 Facsimile of *Daily Sketch*, 14 April 1912. © History of Science Museum, University of Oxford, Marconi Archive Inv. 96891.
37 Cartoon by Leonard Raven-Hill, *Punch*, 22 October 1913. Oxford, Bodleian Library, 2706 d.99, vol. 145, p. 341.
39 *The Wireless Constructor*, November, 1924. Oxford, Bodleian Library, John Johnson Collection: Wireless box 11, p. 69.
45 © History of Science Museum, University of Oxford, Inv. 89558.
47 *The Graphic*, 20 May 1922. Oxford, Bodleian Library, N. 2288 b.7, vol. 105, p. 616.
50 *The Sketch*, 17 October 1923. Oxford, Bodleian Library, N. 17078 c.32, vol. 124, p. 131.

52 *Popular Wireless*, 3 June 1922. Oxford, Bodleian Library, John Johnson Collection: Wireless box 11.
53 *Home Radio* magazine. Oxford, Bodleian Library, John Johnson Collection: Wireless box 11.
54 *The Illustrated London News*, 10 November 1923. Oxford, Bodleian Library, N. 2288 b.6, vol. 163, p. 28.
57 *The Sphere*, 3 May 1924. Oxford, Bodleian Library, N. 2288 b.34, vol. 97, p. 117.
59 © History of Science Museum, University of Oxford, Inv. 50144.
60 *The Illustrated London News*, Christmas Number 1922. Oxford, Bodleian Library, N. 2288 b.6, vol. 161.
64 Cartoon by J.H. Dowd, *Radio Times*, Christmas issue 1925. Oxford, Bodleian Library, Per. 247933 c.10, vol. 9, p. 607.
67 *Oscillation*, p. 13. Oxford, Bodleian Library, John Johnson Collection: Wireless box 5, p. 13. Copyright H. M. Bateman Designs www.hmbateman.com.
73 *Radio Times*, 18 April 1924. Oxford, Bodleian Library, Per. 247933 c.10, vol. 3, p. 168.
74 *The Illustrated London News*, 10 November 1923. Oxford, Bodleian Library, N. 2288 b.6, p. 36.
76 Wireless board game, *c.* 1924. Oxford, Bodleian Library, John Johnson Collection, Ballam Collection: Games 1920s (38).
79 *The Radio Puzzle*, *c.* 1935. Oxford, Bodleian Library, John Johnson Collection, Ballam Collection: Games 1930s (68).
80 Oxford, Bodleian Library, MS. 6241/26, folder 1.
82 Courtesy of Dr Willem Hackmann.
87 *New Ventures in Broadcasting*, 1928, p. 12. Oxford, Bodleian Library, John Johnson Collection: Wireless box 6.
93 *Radio Times*, 9 November 1923. Oxford, Bodleian Library, Per. 247933 c.10, v.1/v.2, p. 234.
95 Oxford, Bodleian Library, John Johnson Collection: Wireless box 1.
97 *Wireless* magazine, 19 September 1925, p. 47. Oxford, Bodleian Library, John Johnson Collection: Wireless box 11.
98 *Radio for the Million*, September 1927. Oxford, Bodleian Library, John Johnson Collection: Wireless box 11.
100 *Radio Times*, 21 May 1926. Oxford, Bodleian Library, Per. 247933 c.10, vol. 11, p. 330.
102 © History of Science Museum, University of Oxford, Inv. 58056.
107 Cartoon in *Punch*, 10 February 1932. Oxford, Bodleian Library, N. 2706 d.10, vol. 182, p. 145.
108 Cartoon in *Radio Times*, 18 July 1926. Oxford, Bodleian Library, Per. 247933 c.10, vol. 12, p. 126.
115 Drawings by John Campbell for Marconi pamphlet. Oxford, Bodleian Library, John Johnson Collection: Wireless box 3.
116 *Radio Times*, 16 November 1934. Oxford, Bodleian Library, Per. 247933 c.10, vol. 45.
118 Illustration by Wanda Radford in *The Graphic*, 2 March 1929. Oxford, Bodleian Library, N. 2288 b.7, vol. 123, p. 306.
120 *Radio Times*, 13 November 1932. Oxford, Bodleian Library, Per. 247933 c.10, vol. 37, p. 419.
123 *Radio Times*, 14 November 1937. Oxford, Bodleian Library, Per. 247933 c.10, vol. 57, p. 7.

125 *Radio Times*, 23 November 1923. Oxford, Bodleian Library, Per. 247933 c.10, vol. 1, p. 316.
127 *Radio Times*, 18 January 1924. Oxford, Bodleian Library, Per. 247933 c.10, vol. 2, p. 123.
129 *Home, Health and Garden*, 1934. Oxford, Bodleian Library, John Johnson Collection: Wireless box 8.
137 Oxford, Bodleian Library, John Johnson Collection: Wireless box 1.
138 *Radio Times*, 23 December 1932, cover design by Edward Ardizzone. Oxford, Bodleian Library, Per. 247933 c.10, vol. 37. © 1932, The Ardizzone Trust; reproduced by permission of David Higham Associates.
141 Chronicle/Alamy.
143 Oxford, Bodleian Library, John Johnson Collection: Wireless box 1.
145 Oxford, Bodleian Library, John Johnson Collection: Horn 23.
149 *The Sketch*, 4 April 1934. Oxford, Bodleian Library, N. 17078 c.32, vol. 166, p. 25.
150 *Broadcasting in Everyday Life*, BBC, 1939. Oxford, Bodleian Library, MSS 6241/50, p. 26.
155 *The Listener*, 1939. Oxford, Bodleian Library, John Johnson Collection: Wireless box 4.
159 Illustration from *Broadcasting House*, 1932, p. 6. Oxford, Bodleian Library, John Johnson Collection: Wireless box 4.
162 Amoret Tanner/Alamy.
164 Oxford, Bodleian Library, Johnson Adds.
169 *Radio Pictorial*, 19 January 1934. Oxford, Bodleian Library, Per. 247934 c.1.
174 Radio Circle certificates designed by Dorothy Hutton. Oxford, Bodleian Library, John Johnson Collection: Wireless box 6.
177 Heath Robinson cartoon in *BBC Yearbook* 1930. Oxford, Bodleian Library, Per. 247933 e.39, p. 386.
180 Oxford, Bodleian Library, John Johnson Collection: Wireless box 8.
181 Oxford, Bodleian Library, John Johnson Collection: Wireless box 8.
182 Oxford, Bodleian Library, John Johnson Collection: Wireless box 8.
188 Still from Mr Murgatroyd and Mr Winterbottom, 1936. British Pathé www.britishpathe.com/asset/79402/.
190 Oxford, Bodleian Library, John Johnson Collection: Wireless box 1.
192 *The Sphere*, 1 February 1936. Oxford, Bodleian Library, N. 2288 b.34, p. 226.
193 Chronicle/Alamy.
203 Oxford, Bodleian Library, John Johnson Collection: Wireless box 2.
203 Gainsborough Films/Album/Alamy.
208 *Radio Times* cover illustration by Kraber (John Rowland Barker), 18 November 1938. Oxford, Bodleian Library, Per. 247933 c.10, vol. 61.
210 Pictorial Press Ltd/Alamy.
212 *Radio Times*, 10 December 1925. Oxford, Bodleian Library, Per. 247933 c.10, vol. 9.
214 Cartoon by Mervyn Wilson, *Radio Times*, 18 November 1938. Oxford, Bodleian Library, Per. 247933 c.10, vol. 61, p. 9.
216 Cartoon by D.L. Ghilchik, *Punch*, 21 November 1923. Oxford, Bodleian Library, N. 2706 d.10, p. 487.
228 Oxford, Bodleian Library, John Johnson Collection: Electricity and Electrical Appliances box 3.
234 *Popular Wireless*, 3 June 1922. Oxford, Bodleian Library, John Johnson Collection: Wireless box 11.

INDEX

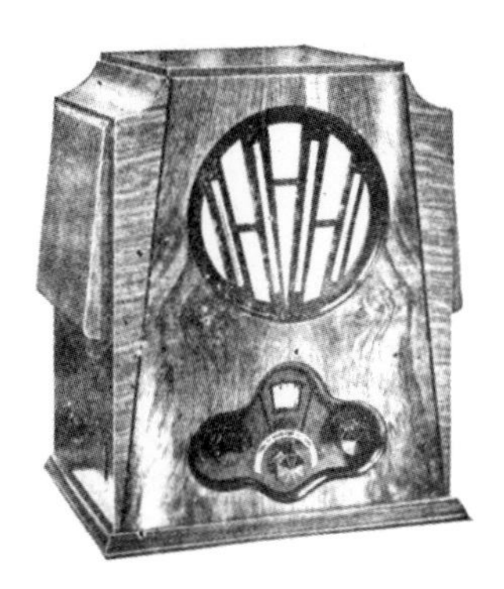